RAND NATIONAL DEFENSE RESEARCH INSTITUTE

Mathematics Audit of the DoDEA Schools

2014–2015

Rita Karam, Brian M. Stecher, Tiffany Tsai, Geoffrey Grimm, Jonathan Schweig

Prepared for the Department of Defense Education Activity

For more information on this publication, visit www.rand.org/t/RR1272

Library of Congress Cataloging-in-Publication Data
ISBN: 978-0-8330-9560-2

Published by the RAND Corporation, Santa Monica, Calif.
© Copyright 2016 RAND Corporation
RAND® is a registered trademark.

Cover: Constantinos/Fotolia

Support RAND
Make a tax-deductible charitable contribution at
www.rand.org/giving/contribute

www.rand.org

Preface

The Department of Defense Education Activity (DoDEA) operates 181 schools to educate military dependents. The schools are located in 11 foreign countries, seven states, Guam, and Puerto Rico. DoDEA is committed to providing a high-quality education for the dependents of military and civilian employees. The Community Strategic Plan for school years 2013–2014 through 2017–2018 commits DoDEA to implement the Common Core State Standards in mathematics (to which DoDEA refers as the College and Career Readiness [CCR] standards). Efforts to change from the 2009 DoDEA mathematics standards to the CCR standards began in earnest during the 2014–2015 school year, with the CCR mathematics standards to be rolled out in elementary schools starting in school year 2015–2016 and scaled up the following school year to middle and high schools. This study was designed to support the transition to the new standards by examining the extent to which DoDEA's existing mathematics program was consistent with best mathematics education practices and by offering suggestions for improving the implementation of the CCR standards in mathematics. Some of the recommendations, such as aligning the CCR standards with new assessments, appear dated because, by the time the report was completed and reviewed, DoDEA had already taken some steps. The study employed a qualitative approach to examine the existing mathematics program and the initial efforts DoDEA was taking to align the mathematics program and practices with the CCR standards in mathematics.

This research was sponsored by the DoDEA and conducted within the Forces and Resources Policy Center of the RAND National Defense Research Institute, a federally funded research and development center sponsored by the Office of the Secretary of Defense, the Joint Staff, the Unified Combatant Commands, the Navy, the Marine Corps, the defense agencies, and the defense Intelligence Community.

For more information on the RAND Forces and Resources Policy Center, see www.rand.org/nsrd/ndri/centers/frp or contact the director (contact information is provided on the web page).

Contents

Figures and Tables

Figures

Tables

Summary

The Department of Defense Education Activity (DoDEA) operates 181 schools to educate military dependents. The schools are located in 11 foreign countries, seven states, Guam, and Puerto Rico. DoDEA's Community Strategic Plan for school years 2013–2014 through 2017–2018 calls for continuously improving the education of military dependents to ensure that they have the skills needed to compete in the 21st-century labor market. The Community Strategic Plan commits DoDEA to implement the Common Core State Standards (CCSS), to which DoDEA refers as the College and Career Readiness (CCR) standards, as a way to ensure the continued provision of high-quality education to military dependents. This study examined the quality of the current DoDEA kindergarten through 12th grade mathematics program (which is focused on mathematics standards adopted in 2009) in eight key areas identified by DoDEA. We also examined the extent to which teachers' and administrators' current practices supported the implementation of the CCR standards in mathematics, identified concerns related to the implementation of the CCR standards, and made recommendations for improving implementation (which began during the latter part of our study) in the future.

In preparation for conducting the study, we reviewed the research literature to identify best practices in standards-based mathematics instruction in the areas of interest to DoDEA. To examine effective strategies for implementing the CCR standards, we also interviewed administrators in three exemplary local education agencies (LEAs) selected by DoDEA for this purpose.[1] We visited 25 schools located in nine of the 14 DoDEA school districts, including schools in all three regions—the Americas, Europe, and the Pacific. Schools were selected purposively by DoDEA to represent different geographic areas, grade spans, and school sizes. We spent one full day in each school; during this time we interviewed the principal and two teachers and observed two mathematics lessons. To obtain similar information from the remaining five districts that were not visited, we conducted telephone interviews and focus groups with superintendents, mathematics Instructional Support Specialists (ISSs) and school principals. We also interviewed administrative staff at DoDEA headquarters (HQ) and at the three area offices. In addition, we collected information about teachers' needs for support through an anonymous online survey made available to all DoDEA mathematics teachers.

It is worth noting that the study was conducted at the same time that DoDEA was beginning the implementation of the CCR standards in mathematics, and some of our findings may be out of date by the time this report is published.

Implementing a high-quality mathematics program that incorporates best practices in each of the eight key DoDEA areas is a challenging endeavor. Adoption of the CCR standards

---

[1] DoDEA identified five LEAs it believed had exemplary mathematics programs; we included them in the study based on willingness to participate.

is one step toward improving the program in the future. As a basis for that change, our results suggest that DoDEA has been successful in incorporating some best practices into its existing mathematics program based on 2009 standards more than others. In particular, we observed more best practices in the areas of Equity and Diversity and Classroom Assessment than in the other six topic areas.

Curriculum Resources

Many interviewed teachers indicated that their textbooks were not well aligned with the 2009 DoDEA mathematics standards or did not provide sufficient lesson alternatives to meet the needs of all of their students. These teachers reported not having sufficient standards-aligned curriculum resources readily available, and, as a result, some sought additional resources on their own to incorporate into their lessons. Many interviewed teachers were also concerned that there was no system in place to help them identify and assess the quality of the alternative resources they identified.

Curriculum and Instructional Quality

The literature emphasizes three aspects of high-quality mathematics curriculum and instruction: (1) focus on fewer topics but in greater depth; (2) coherence among mathematics topics across grades and subject areas; and (3) rigor of mathematical thinking, placing equal emphasis on conceptual understanding and procedural fluency. The DoDEA mathematics program we evaluated had a large number of standards that made the program challenging for teachers to master, thus affecting the implementation of the mathematics program. Furthermore, the extent to which lessons embodied coherence and rigor varied. In most lessons we observed, teachers made connections between mathematical concepts within the lesson but did not connect to other topics; in addition, they did not explore the ideas in depth or provide opportunities for students to discuss them. Connections between mathematics and other subjects were infrequent. Although most teachers engaged in activities to promote understanding, procedural skills, and applications, balancing the three areas was challenging in more than half of the classrooms observed. Teachers seemed to be more comfortable emphasizing mathematical procedures. Nevertheless, some lessons we observed incorporated challenging activities designed to promote deeper understanding of mathematical concepts. These activities addressed authentic settings that required students to make assumptions and predict future outcomes.

Equity and Diversity

All schools visited had systems in place to examine student performance and identify low-performing students. Schools also had effective systems for identifying English learners and students with special needs, but the identification of gifted students was more informal. Almost all schools visited implemented inclusion policies, but such policies were not accompanied by differentiation in instruction in the classroom to ensure student access to the curriculum.

Content Knowledge

We did not assess teachers' mathematics content knowledge directly, but we did examine the extent to which teachers were familiar with the mathematics content standards that were in place at the time of the study and the CCR content standards that were slated for implementation. There was a widespread perception among the teachers and administrators we interviewed that teachers were familiar with the content of the 2009 DoDEA standards, but they were not familiar with the CCR standards. There were some concerns that teachers, especially at the elementary grade levels, lacked either in-depth content knowledge of mathematics or confidence in their knowledge.

Classroom Assessments

Teachers used a wide range of assessment techniques to monitor student progress. Although the frequency and type of classroom assessment varied considerably across teachers, content-focused assessment and assessments measuring procedural fluency were more prevalent than assessments that tried to measure mathematical practices (e.g., processes and proficiencies). Teachers made efforts to prepare students for the summative tests students were required to take, and that influenced the ways they assessed students within their classrooms. Teachers indicated that they would need training on assessment that is aligned with the CCR standards when these standards are adopted.

Instructional Leadership

DoDEA HQ has a strong voice in setting the instructional direction in mathematics for schools; a number of administrators reported that their role was to carry out directions from HQ. On a day-to-day basis, it appeared to us that principals and teachers shared responsibility for instructional leadership in mathematics. However, principals did not always self-identify—and were not always identified by teachers—as the instructional leader in mathematics. In many schools, mathematics instructional leadership fell informally to other individuals, including mathematics ISSs, department chairs, or individual teachers, depending on staff capacities and expertise.

Administrator and Teacher Professional Development

Principals and teachers have received very little professional development (PD) pertaining to mathematics since the adoption of the DoDEA standards in 2009. Most reported a need for PD to prepare them for the CCR standards and, to a lesser extent, to understand specific mathematics topics.

xii Mathematics Audit of the DoDEA Schools: 2014–2015

Science, Technology, Engineering, and Mathematics Opportunities

Most schools provided some science, technology, engineering, and mathematics (STEM) enrichment opportunities to their students, including STEM clubs, STEM events (e.g., science fairs), and STEM activities integrated into the school day (e.g., STEM lab). However, student access to STEM opportunities varied considerably depending on staff expertise, resources, and proximity to STEM-rich organizations.

Recommendations

Move Quickly to Align Mathematics Curriculum, Assessments, and Support Services with the College and Career Readiness Standards

There are a number of steps that need to be taken to align all elements of the mathematics program with the CCR standards, and DoDEA is well aware of the challenges it faces to align textbooks, assessments, and continuing training and support with the CCR standards. All three need to be replaced or transformed, and each step is expensive. One key intermediate step might be to provide more CCR standards–aligned curriculum resources to bridge the time before adoption of new textbooks and assessments. These resources could include a scope-and-sequence document or a pacing guide to help teachers match lesson content from current resources with grade-level standards in the CCR standards.

Provide Training to Teachers on Locating Online Resources

Many of the teachers we interviewed indicated that they were using online resources to supplement the textbooks and other materials available from DoDEA. As U.S. states and districts move to implement the CCSS, more resources are being made available online. All DoDEA teachers should be able to access such resources and should receive training on how to judge the quality and applicability of digital resources to meet their own instructional needs.

Help Districts Develop Messages for Parents

The three LEAs we interviewed provided parent workshops as part of their CCSS implementation efforts. They reported that it was important to keep parents informed about curricular changes, and it was critical to offer them opportunities to familiarize themselves with the skills their children would be asked to master and the kinds of lessons their children would be receiving. Rather than expecting each school to create its own parent education program, it would benefit schools and parents if DoDEA were to take the lead in preparing materials for parents and developing workshops to help schools communicate about the CCR standards in mathematics.

Support Teachers' Efforts to Change Practice by Reducing Test-Based Accountability Pressures

Many teachers reported to us that they felt pressure to prepare students to do well on the TerraNova, the Preliminary SAT/National Merit Scholarship Qualifying Test (PSAT/NMSQT), and the SAT because results from these tests were used as indicators of school progress. Some teachers told us that one of their concerns about the CCR standards is that teaching to the new standards would not prepare their students adequately for these standardized tests.

Under these circumstances, DoDEA might want to consider ways to deflect these pressures, particularly if there are plans to change assessments in the future. For example, it might be helpful to share examples of questions from CCSS-aligned assessments, such as Smarter Balanced or the Partnership for Assessment of Readiness for College and Careers, to help teachers understand the kinds of expectations DoDEA has for student understanding in mathematics.

Provide Time and Resources for High-Quality Professional Development for Mathematics Instructional Support Specialists, Principals, and Teachers

DoDEA understands the importance of preparing staff to implement the CCR standards, and it is already making efforts to provide extensive training for key staff at all levels of the system. At least on the surface, the plans that have been described to us seem to address most of the concerns that we heard from administrators and staff. We would like to point out two or three particular issues that might warrant attention related to PD.

Sustain the Training for Two or Three Years

The exemplary LEAs whose representatives we interviewed found it necessary to continue to support teachers for three years to change their instruction to align with the CCR standards.[2] We do not know DoDEA's long-term strategy for PD related to aligning mathematics to the CCR standards, but we encourage leadership to sustain the support provided to teachers for a minimum of two years and suggest that three years may be appropriate.

Provide Resources and Support at the School and Classroom Levels

Teaching is a highly "situated" activity, i.e., it depends a great deal on the conditions in a given school, the skills and personality of a given teacher, and the set of students and resources assembled in a specific classroom. Thus, it is hard to improve teaching from a distance. We recommend that DoDEA find ways to bring additional instructional support directly to teachers. This might be accomplished in many ways, but, through whatever means selected, DoDEA should strive to enhance individualized and classroom-situated support services for teachers.

Build School Capacity

While centralized DoDEA support for the CCR standards is vital, no reform will be successful unless it brings about sustainable changes at the school level. For this to happen, we recommend that schools have an assigned individual or staff person who is knowledgeable about the CCR content standards and practice standards and adept at supporting teachers in lesson development. DoDEA might want to consider supporting a mathematics specialist or mathematics teacher-leader, who will work closely with district ISSs in the provision of teacher support. Principals and teachers in schools with mathematics specialists on staff were very positive about the type and quality of support provided by the specialists. In addition, having active and reliable support in the form of professional learning communities may help to develop and maintain enthusiasm for the CCR standards at the school, even as leadership or other staff changes.

---

[2] The LEAs referred to the standards as the CCSS.

Start Working with High Schools Now to Ensure Broad Support for Future College and Career Readiness Standards Implementation

At the time of our visits, most high school teachers indicated that they had not heard much about the CCR standards other than that DoDEA had adopted them. Moreover, high school teachers were concerned about how the CCR standards would affect their students' college preparation and their overall mathematics program. We suggest that DoDEA send clearer messages about its intentions, particularly at the high-school level, to promote buy-in and support prior to implementation.

Monitor College and Career Readiness Standards Implementation and Outcomes

Previous research on school reform shows that the level and quality of implementation are major determinants of outcomes. Hence, it is important to monitor changes in practices and strategies undertaken by schools and teachers in order to identify challenges and act to overcome them. We recommend that DoDEA consider establishing a system to regularly collect information on school practices as well as on academic and nonacademic student outcomes (e.g., student scores on CCR standards–aligned assessments, student motivation, interest in mathematics). This system would provide information to DoDEA so that it can monitor the CCR standards implementation, identify areas in need of improvement at the system level, and provide timely and targeted support.

Next Steps

In addition to these specific recommendations, we also identified three questions that DoDEA might want to investigate as it continues to implement the CCR standards in the future. Answers to these questions will help DoDEA determine its next steps to support full implementation of the new standards. The questions are: How effective is the 2015–2016 training and support in preparing teachers to implement the CCR standards in mathematics? What follow-on efforts are needed to support implementation in subsequent years? Does the adoption of the CCR standards and accompanying changes in curriculum and instruction lead to better student outcomes?

Examining these questions in depth not only will improve the CCR standards implementation and outcomes in mathematics, it also will inform planned efforts to implement the CCR standards in English language arts in the coming year.

Acknowledgments

In the course of this study, we visited dozens of Department of Defense Education Activity (DoDEA) schools and spoke with hundreds of staff, and we could not have completed this study without their courteous and candid participation. Special thanks are due to the administrators and support staff in selected DoDEA districts, who helped us schedule visits, provided logistical support, and, in some cases, acted as guide or chauffeur to facilitate our local visits. Lisa Holloway and Sarah Koebley guided the effort from the outset, helping us frame the study, clarify DoDEA priorities, obtain access to people and places, and generally keep us on target in a rapidly changing context. Our two reviewers, Catherine Augustine and Anna Saavedra, provided valuable feedback and made suggestions that improved the clarity and quality of the final report.

Abbreviations

ASCD	Association for Supervision and Curriculum Development
CCR	College and Career Readiness
CCSS	Common Core State Standards
CCSSO	Council of Chief State School Officers
CSI	continuous school improvement
CSP	Community Strategic Plan
DoDEA	Department of Defense Education Activity
EL	English learner
ELA	English-language arts
ESL	English as a second language
HQ	DoDEA headquarters
ISS	Instructional Support Specialist (district level)
LEA	local education agency
K–12	kindergarten through 12th grade
MKT	Mathematical Knowledge for Teaching
NAESP	National Association of Elementary School Principals
NCTM	National Council of Teachers of Mathematics
PARCC	Partnership for Assessment of Readiness for College and Careers
PD	professional development
pre-K	prekindergarten
PSAT/NMSQT	Preliminary SAT/National Merit Scholarship Qualifying Test
SST	student support team
STEM	science, technology, engineering, and mathematics

CHAPTER ONE

Introduction

Background

The Department of Defense Education Activity (DoDEA) operates 181 schools in 14 districts located in 11 foreign countries, seven states, Guam, and Puerto Rico. It currently serves more than 78,000 students and employs more than 8,000 teachers (Clark, 2015). DoDEA is committed to providing a high-quality education for the dependents of military and civilian employees. The Community Strategic Plan (CSP) for school years 2013–2014 through 2017–2018 calls for DoDEA to continuously improve its efforts to educate students to excel in an "increasingly competitive, global 21st Century environment" (DoDEA, undated). Toward that end, the CSP commits DoDEA to implement the Common Core State Standards (CCSS), to which DoDEA refers as the College and Career Readiness (CCR) standards, in mathematics and English-language arts (ELA).[1] This report focused on the mathematics program.

DoDEA sees a number of benefits to implementing the CCR standards in mathematics for military-connected students. Although DoDEA students have scored above the U.S. national average on the National Assessment of Educational Progress, their performance on international comparisons of mathematics, reading, and science competencies was low compared with that of students in other developed countries (Organization for Economic Co-operation and Development [OECD], 2011). DoDEA administrators recognized the need for improvement and have identified adoption of the CCR standards as an important strategy for raising academic standards and achievement. Further, children of military families often move between state public schools and those in the DoDEA system, changing schools as many as six times during the course of their school years (Richmond, 2015). Since most states have adopted the CCSS, DoDEA students would be better prepared to transfer to state public schools if they were educated to master the new standards. To ensure a smooth transition between the previous DoDEA standards (adopted in 2009) and the CCR standards, DoDEA plans to spend the next three to five years phasing in the CCR standards, starting with mathematics in prekindergarten (pre-K) through 5th grade, then expanding to secondary mathematics and to ELA.

DoDEA is taking steps toward aligning its mathematics program to the CCR standards. It recognizes that the CCR standards entail a number of changes from the mathematics standards used in DoDEA schools since 2009. For example, the CCR standards grade-level content and the sequence in which topics in mathematics are introduced in the elementary

---

[1] In this document, we refer to the new DoDEA standards as the CCR standards, the label currently used within DoDEA. We refer to the standards adopted by the majority of the states as the CCSS. These two sets of standards are essentially the same.

grades differ from that contained in the 2009 standards.[2] In addition to differences in the grade-level content, the CCR's "Standards for Mathematical Practice" identify eight types of mathematical expertise that teachers should seek to develop in their students, ranging from "mak[ing] sense of problems and persevering in solving them" to "attend[ing] to precision" (Common Core State Standards Initiative, undated [b]). While the 2009 DoDEA mathematics standards delineate both content and process expectations, they do not align exactly with the content and practice standards in the CCR standards (Common Core State Standards Initiative, undated [b]). Thus, the adoption of the CCR standards entails changes in the content being taught (e.g., topics addressed), when mathematical topics are introduced, and what mathematical practices are emphasized.

New standards by themselves are not sufficient to ensure that students will receive excellent education in mathematics. The mathematics standards establish the formal expectations of the DoDEA mathematics program, and all other aspects of the program need to be oriented to achieve those expectations. For example, curriculum materials have to be adjusted to align with the standards; classroom assessments need to be modified to match the demands of the standards; lesson plans need to be adapted to emphasize the important connections among mathematical ideas; and instructional interactions need to change to foster perseverance in problem-solving, attention to precision, among others. Furthermore, as noted in the CSP, schools also need to increase curricular and cocurricular opportunities to build student competencies in science, technology, engineering, and mathematics (STEM) fields.

Purpose and Approach

DoDEA conducted an analysis to identify the gaps between the content and performance expectations in the 2009 DoDEA mathematics standards and the CCR standards. The analysis found that the number of standards and depth of knowledge embodied in the standards were similar, but there were many ways in which they differed. As the authors summarized,

> Across the grades, there are many partial alignments because the specific mathematical expectations for conceptual understanding, specified levels of procedural skill and fluency, methods of representation (e.g., double number lines or ratio tables), and reference to applications (real-world or mathematical problems) in the Common Core standards were not found in the DoDEA standards. (Johnston et al., undated, p. 6)

They also found high alignment between the current DoDEA textbooks and the CCR standards in kindergarten and first grade but noted that the alignment declined progressively in higher grades.

The successful adoption of the CCR standards requires changes not only in mathematics content, but also in supporting actions in other critical areas of the teaching and learning processes. In this context DoDEA asked the RAND National Defense Research Institute to examine the quality of the current mathematics program and identify program strengths and weaknesses in eight key areas:

---

[2] DoDEA commissioned an analysis of gaps between mathematics standards (adopted in 2009) and the performance expectations of the CCSS. See Johnston et al., undated.

1. curriculum resources
2. curriculum and instructional quality
3. equity and diversity
4. teacher content knowledge[3]
5. curriculum assessments
6. instructional leadership
7. administrator and teacher professional development (PD)
8. access to STEM enrichment.

For these areas, the study examines (1) the quality of the practices, curriculum materials, processes, and supplemental mathematics products currently used in the mathematics program and (2) the capacity of DoDEA to support the implementation of the CCR standards. We reviewed research on effective implementation of the CCSS and other curriculum standards as well as research on mathematics education, to identify best practices associated with each of the eight key areas of the mathematics program. We also interviewed administrators in three local education agencies (LEAs) identified by DoDEA as exemplary in terms of implementing the CCSS in mathematics and used them as examples for comparing DoDEA efforts. As explained in more detail in this report, we use a combination of interviews, classroom observations, and teacher surveys to examine current DoDEA practices in relation to the best practices identified through the literature and exemplary districts.

Organization of the Report

In Chapter Two, we describe the methodology used to assess the quality of the current mathematics program, and we provide a conceptual framework for the study based on the research literature on best practices. Chapter Three presents findings on the quality of each of the eight key mathematics program areas and examines how aligned current DoDEA mathematics practices are with best practices associated with the CCSS. Chapter Four describes insights we gained about implementation of curriculum reforms in DoDEA. The final chapter provides recommendations on promoting quality teaching and learning in mathematics aligned with the CCR standards and scaling up the implementation of reform effort to middle and high schools. It also suggests unanswered questions that might guide future research to help DoDEA take next steps in implementation.

---

[3] We do not directly test teachers' content knowledge; instead, we rely on anecdotal information, primarily based on principal and teacher interviews.

CHAPTER TWO

Methods and Review of Best Practices

In this chapter, we provide details on the study methods, including data collection and analysis. We then review the literature on best practices for standards-based mathematics programs and effective standards implementation. We discuss findings from interviews with three LEAs that serve as best-case examples for comparing DoDEA efforts in implementing the CCR standards, which we incorporate into our notion of best practices. Finally, we summarize the key indicators for judging the quality of DoDEA's current mathematics program and assessing DoDEA's effort as it strives to implement the CCR standards.

Methods

Qualitative Approach

We chose a primarily qualitative case-study approach for this project. The phenomena we were studying were complex, unfolding in real time, and involved a wide range of stakeholders across the DoDEA system. Specifically, the study was concerned with examining how various educational stakeholders in the system were implementing the 2009 DoDEA mathematics standards and the steps they were taking to align their mathematics program with the CCR standards. Measuring implementation requires paying attention to details and accounting for what happens as individuals throughout the system act on the design of the program. Researchers have found that the best way to gather relevant information in contexts such as this is to perform in-depth case studies, which combine school and classroom observations with face-to-face interviews with individuals across the system who are involved directly in the implementation process, from district administrators and staff to principals to teachers. We visited a substantial number of DoDEA districts and schools, but it was not feasible to visit them all (see the next section). Because of this limitation, and based on a request from one of the teacher unions, we added to an online teacher survey to provide teachers with the opportunity to share their experiences with the mathematics program and the preparation they had for adopting the CCR standards.

Sampling of DoDEA Stakeholders

Since it was not feasible to visit all DoDEA schools or interview all administrators and teachers, we used a two-stage sampling process to obtain as representative a group of respondents as possible. In the first stage of sampling, DoDEA identified three districts in each of DoDEA's three geographical areas—the Americas, Europe, and the Pacific—and selected two or three schools in each of these nine districts (see Appendix A for a list of the schools visited). Schools were selected to vary in terms of school size, grade span, and geographic location. The first-stage sample included a mixture of primary schools, elementary schools, middle/intermediate schools, and high schools. We conducted case studies, including interviews and observations in these 25 schools. Overall, the demographic characteristics of the 25 schools that were visited for this study are similar to those of the rest of the DoDEA schools (see Appendix B for a detailed breakdown of school characteristics). At all educational levels, the schools had an even breakdown of males and females, with slightly higher percentages of males than females. In terms of race and ethnicity, there were almost as many white students as students from other ethnic backgrounds. The major difference between our sample and the rest of the DoDEA schools was the percentage of English learners (ELs). At the elementary and middle school levels, DoDEA schools that were visited had higher percentages of ELs than the rest of the DoDEA schools. This difference could be the result of selecting Puerto Rico as one of the districts, since it has an unusually high percentage of ELs. Mathematics proficiency levels were also similar to those of the rest of DoDEA schools.

In the second stage of sampling, for these nine districts and 25 schools, we obtained contact information for the district superintendent, the principal in each school, and the mathematics Instructional Support Specialist (ISS); we also obtained lists of all teachers who taught mathematics. We asked the district superintendent to notify the principals about the study, then we contacted the principals to identify specific teachers for participation and to arrange site visits. We asked the principal to identify teachers who were most familiar with the mathematics program in the school and who would be able to provide information on implementation issues that they and their peers face. These teachers, considered "leaders" by their principals, were our first candidates for interviews. We also randomly selected two teachers as candidates for observations, attempting to obtain as much variation as possible across grade levels and subjects. To reduce burden on teachers and to capture as much variation in teacher experience as possible, we tried not to interview and observe the same teachers. We contacted each teacher directly and invited him or her to participate in the study, indicating that participation was voluntary. If a teacher declined to participate, we replaced him or her with another teacher from the school. In some cases, we were not able to find four teachers to participate, so we asked volunteers to participate in both the observation and interview. In some cases, we could not conduct all the planned interviews or observations (e.g., absences, field trips, small schools). We also asked to interview the mathematics ISS if that person was available at the school.

In addition, we conducted telephone focus groups with superintendents, principals, and mathematics ISSs from the remaining five DoDEA districts that we did not visit. All superintendents and ISSs from these five districts were invited to participate in separate focus-group conference calls, and we talked with those who opted to participate. A sample of five principals from the five districts was selected randomly; we invited them to participate in a focus-group telephone call, and we spoke with those who agreed to be involved. We also interviewed directors or deputy directors of DoDEA area offices.

Finally, at DoDEA's request, we developed an online survey that teachers could complete anonymously. This allowed us to gather additional data from greater numbers of teachers within the DoDEA system. DoDEA sent the link to the online survey to mathematics teachers by electronic mail, and teachers decided on their own whether to complete the survey. A total of 699 teachers completed the online survey—54 percent were elementary school teachers, and 46 percent were secondary school teachers. Fifty-one percent of the respondents were teachers located in Europe, 32 percent were located in the Americas, and 17 percent were located in the Pacific. This represents about 8 percent of all DoDEA teachers; because we did not have a full count of mathematic teachers, we were unable to determine what percentage of mathematics teachers the total completions constituted.

Table 2.1 shows the total number of interview participants by stakeholder group for visited and nonvisited districts.

Data Collection

We collected the following data between February 2015 and April 2015.

Site Visits (Including Interviews, Observations, and Lesson Artifacts)

We visited nine districts and 25 schools in the Americas, Europe, and the Pacific. The schools were selected by DoDEA to represent a variety of geographic locations, student population demographics, school levels, and proximity to district offices. The objectives of these visits were to capture varying experiences in implementing the mathematics program and to obtain a wide range of perspectives about DoDEA's unfolding efforts to implement the CCR standards. Teams of two RAND researchers spent one day at each school. During each visit, we met individually with the school principal and two mathematics teachers or instructional leaders identified by the school principal, and we observed two mathematics lessons. In some cases, we also interviewed the district superintendent and the district mathematics ISS during this visit, if scheduling permitted. Otherwise, we conducted phone interviews with these individuals after we completed the site visit.

We developed a set of interview protocols to gather information about the eight key mathematics program features identified by DoDEA: (1) curriculum resources, (2) curriculum

Table 2.1
Number of Interviewees, by Stakeholder Group

Source	Group	Participants
Visited DoDEA districts (9)	Superintendents	8
	ISSs	12
	Principals	25
	Teachers	48
	Classroom observations	47
Non-visited DoDEA districts (5)	Superintendents	4 (two focus groups)
	ISSs	5 (2 focus groups)
	State officials	5
	Principals	2 (1 focus group)
DoDEA area office	Director and deputy directors	4
Online survey (DoDEA)	Mathematics teachers	699

and instructional quality, (3) equity and diversity, (4) teacher content knowledge, (5) classroom assessment, (6) instructional leadership, (7) PD in mathematics, and (8) STEM opportunities. To cover these topics, but also allow for new features to emerge, we used semistructured interviews that included open-ended questions with supplemental probes to examine specific topics. The protocols were common across the sites but differed slightly for each type of interviewee—principals, teachers, ISSs, and superintendents. For example, we asked teachers about instructional practices commonly used in their classrooms and the PD they received. On the other hand, we asked superintendents for their broader perspective on DoDEA's current move toward implementing the CCR standards, curriculum resources, instructional quality, and efforts to improve mathematics instruction.

We also conducted two classroom observations in each school and asked each teacher to provide artifacts related to the lessons. Later in the day, we rated each lesson and each set of artifacts using a rubric developed with feedback from DoDEA. The rubric covered 11 dimensions of quality mathematics instruction, including (1) clarity of lesson objectives, (2) lesson structure and coherence, (3) student explanations, (4) connections between mathematics concepts, (5) connections between different disciplines, (6) cognitive challenge, (7) modeling with mathematics, (8) responsiveness to diverse student needs, (9) appropriate use of tools, (10) student engagement with mathematical content, and (11) formative assessment. For each dimension, we defined what is considered low-, middle-, and high-level implementation (see Appendix C for a copy of the rubric). The rubric allowed us to identify instructional strengths and areas in need of improvement that DoDEA might want to address as it moves toward implementing new standards.

In addition, in each class we observed, we collected lesson materials and student work to help analyze the quality of the mathematics program. These artifacts included lesson plans and prepared slides (for the day observed and one additional day during the same unit), a recently administered homework assignment with student work and answer key or scoring rubric, and a formal classroom-assessment task with student work and answer key or scoring rubric. The availability of the artifacts varied across sites and among teachers. We were unable to obtain the full list of classroom artifacts from every teacher we observed.

During each classroom observation, two RAND researchers took detailed handwritten notes guided by the rubric dimension. After each observation, RAND researchers discussed the notes and rated each dimension on a scale that ranged from one (low) to three (high). Observers also recorded written justifications for each rating, which permitted subsequent data cleaning to improve consistency across raters (see Appendix D for a summary of observation ratings). RAND researchers implemented a similar process for the review of the artifacts. For each teacher, the artifacts were analyzed and rated using the rubric dimensions and corresponding definitions, with a detailed justification regarding the rating (see Appendix D for a summary of the artifact ratings).

Telephone Interviews with Area-Level Staff, Districts, and Schools Not Visited

For the remaining five districts that were not visited, we conducted two telephone focus-group interviews with superintendents, two telephone focus groups with mathematics ISSs, and one telephone focus group with principals. In addition, we conducted individual telephone interviews with area-level directors; deputy directors of curriculum, instruction, and assessment; and area-level mathematics ISSs. We used abbreviated versions of the corresponding site-visit interview protocols during these focus groups. The focus groups elicited information on

DoDEA's efforts to implement the CCR standards, curriculum resources, instructional quality and efforts to improve mathematics instruction, the causes of student success in mathematics, and the gaps in mathematics teaching and learning.

Anonymous Online Teacher Survey

Prompted by a request from one of DoDEA's teacher unions, we also developed a short online survey to ensure that all mathematics teachers in DoDEA had the opportunity to provide general information about the mathematics program. To reduce the burden, we focused the survey on a subset of topics from the interviews, and, to ensure anonymity, we made participation voluntary.

Some of the online survey questions were modified from existing RAND online survey tools, such as the American Teacher Panel (RAND Education, undated). The survey asked teachers about their familiarity with the CCR standards, the instructional practices they generally emphasized during classroom instructions, and the frequency of use of various strategies. The survey also asked about the PD teachers received related to mathematics and their needs for training. The survey concluded with an open-ended question about teachers' perceived strengths and weaknesses of the mathematics program; 375 teachers provided open-ended comments. (See Appendix F for a tabulation of survey results.) Because the response rate for the survey was relatively low, we placed less emphasis on reporting findings from the survey than on those from the interviews and observations when examining the quality of instructional practices.

Exemplary LEA Interviews

At the beginning of the study, we interviewed administrators in three LEAs—Cambridge Public Schools, Oceanside Unified School District, and East Lansing Public Schools—drawn from a list of five districts recommended by DoDEA. We selected the first three districts that agreed to participate in the study. DoDEA considered these LEAs exemplary because of the quality of their mathematics programs and strength of their graduation standards. Some also serve a large number of military-connected students, and thus have experience with the challenges that face this student population. We interviewed associate superintendents or directors of curriculum and instruction, who are leading the CCSS implementation, to gain comparative perspective that would inform best practices. In the interviews, we asked questions about districts' efforts to implement the CCSS in mathematics, the challenges they faced and how they were able to overcome them, and the lessons learned that could inform DoDEA's efforts. Their experiences helped us formulate our best-practice standards.

Data Analysis

We analyzed the interview notes and the observation and artifact ratings to examine the quality of the current mathematics program in each of the eight key areas. We also incorporated results from the online teacher survey where relevant.

To do this, we created specific indicators from the literature (see "Literature Review") and findings from the LEA interviews (see "Local Education Agencies' Lessons Learned"). We then identified practices from the DoDEA interviews, as well as classroom observation and artifact ratings, and organized the identified practices within each of the eight key areas being evaluated across each site (district and corresponding schools visited). For each site, we then compared the practices in each of the eight key areas with the best-practice indicators in mathematics we derived from the literature and the LEA lessons learned. The interview data were then analyzed for cross-site patterns to address common themes and lessons learned. When discussing interview results, we use the term *most* to indicate that more than 50 percent of the respondents who were asked the question had comparable responses or views. Similarly, the term *some* represents 20 to 50 percent of respondents having similar views, and the term *few* reflects less than 20 percent of respondents.

Variation among the sites provided us with the means to draw interesting contrasts that could help educators and policymakers understand how contextual differences might affect the implementation of a high-quality mathematics program.

We tabulated teacher survey responses by region and school level to examine any differences in practices. As indicated earlier, survey data were not used as a primary source for examining quality because we do not know the representativeness of the survey respondents, but, when applicable, we highlighted differences in teacher survey reports; face-to-face interviews; classroom observations, especially in the areas of curriculum and instructional quality; and classroom assessments. We also used teacher surveys to complement findings from DoDEA stakeholder interviews (e.g., superintendents, principals, teachers) regarding teachers' needs for and receipt of PD. We also analyzed the major themes that appeared in the open-ended survey responses and incorporated these into our findings.

Study Limitations

The study collected information from all DoDEA districts and conducted in-depth examination of 25 of its 181 schools. The school sample appears to be diverse and representative, including schools from nine districts in all three DoDEA areas and including primary, elementary, middle, and high schools. While the schools did not appear to us to be atypical in any way, the case-study findings reflect this specific set of schools. We identified and interviewed school personnel (e.g., math instructional leaders, local math support specialists) who are knowledgeable of how the math program is implemented at their schools. In addition, we randomly selected teachers within these 25 schools for classroom observations. Although much of the qualitative study data were self-reported by teachers, we tried to enhance the validity of our findings by observing teacher behavior in classrooms and by collecting and analyzing instructional artifacts. Obtaining data on quality from multiple sources is a method commonly used to obtain reliable measures of complex constructs. We visited each school for one day, and it is possible schools made special preparations, so we might not have seen typical practice. Furthermore, the day we observed classes and interviewed teachers might not be typical of the full year. Nevertheless, the fact that common themes regarding practices emerged from various data sources and across sites (both from sites that we visited at different times and from our phone interviews with districts that were not visited) increases our confidence in the results. Finally, while all teachers were offered the opportunity to complete the online survey, only a small percentage of all teachers participated. Thus, while the online survey results were informative, they are not representative of DoDEA mathematics teachers and should be interpreted with care.

Literature Review

Several literatures were found to be relevant to this study: literature in the field of mathematics education, including work by Deborah Ball and Marilyn Burns; literature on effective school leadership and standards implementation; and literature specific to the implementation of the CCSS. We developed best-practice indicators for each dimension of interest through our review of these literatures.

Curriculum Resources

The availability of standards-aligned instructional materials is critical to the success of standards-based education (Krajcik, McNeill, and Reiser, 2008; Hamilton, Stecher, and Yuan, 2008; Knapp, 1997). Some research has noted that district and school implementation of standards-based instruction can be seriously compromised if standards-aligned curriculum resources are not readily available (Kendall, 2011). On the other hand, curricular resources in and of themselves have been shown to have limited influence on teachers' instructional practices (Fullan, 1991; Ball and Cohen, 1996; Coburn, 2001). This occurs in large part because of the wide variation in how teachers choose to use these resources and because of differences between what is referred to as the *intended* curriculum and the *enacted* curriculum (i.e., how teachers implement curricular resources in practice) (Charalambous, 2010; Stein, Grover, and Henningsen, 1996). In fact, the implementation of a curriculum has been shown to be an important factor in student achievement (Schoen et al., 2003), while fidelity of implementation of standards-based curriculum is equally critical to developing mathematical proficiency and increasing student achievement (Boston and Smith, 2009; Balfanz, Mac Iver, and Byrnes, 2006). High-quality curriculum resources can provide "probabilistic opportunities to influence student thinking" (Charalambous, 2010, p. 249), but providing teachers with support and guidance on how to implement those materials in classrooms to support standards-aligned instruction is a critical step in ensuring that the curriculum resources are used to reach productive instructional ends (Davis and Krajcik, 2005; Confrey and Krupa, 2010; Stein and Kaufman, 2010). These supports can include ongoing PD (Balfanz, Mac Iver, and Byrnes, 2006), as well as curriculum frameworks that encourage educators to identify, develop, and try out standards-aligned materials and evaluate their alignment with standards (Achieve and Education First, 2012).

Curriculum and Instructional Quality

Historically, standards-based educational reform in the United States has resulted in the creation of documents that contain a large number of tightly prescriptive standards (Kendall, 2011; Alberti, 2013; Schmidt, Wang, and McKnight, 2005). Accordingly, curricula in U.S. schools have traditionally been described as "a mile wide and an inch deep" (Schmidt, McKnight, and Raizen, 1997) and in which students are exposed to many topics but spend little time immersed in any particular mathematical idea or concept (Alberti, 2013). However, international comparative research has shown that countries with strong mathematics education programs tended to use curricula that share three key features: (1) *focus* (a small number of topics are covered in great depth), (2) *coherence* (major mathematical topics are linked within and across grades), and (3) *rigor* (conceptual understanding and procedure are given equal attention) (Schmidt, Wang, and McKnight, 2005; Schmidt, McKnight, and Houang, 2001; Schmidt and Houang, 2012). The National Council of Teachers of Mathematics (NCTM) has also noted that these particular features are central to framing discussions about standards and building a high-quality curricula (NCTM, 2000), and next-generation mathematics standards, such as the CCSS for mathematics, have been designed with these principles in mind (Achieve and Education First, 2012; Alberti, 2013).

There is a large body of scholarly work that defines mathematical proficiency and illuminates the aspects of teaching and learning that promote the development of this proficiency (Kilpatrick, Swafford, and Findell, 2001; Hiebert, 2003). NCTM and the National Research Council synthesized this work over the years and used it as the basis for developing the standards for mathematical practice that are a component of the CCSS (Common Core State Standards Initiative, undated [b]). First, high-quality instruction *promotes fluency with procedures* (National Research Council, 2001) and involves the flexibility and accuracy with which students can apply mathematical knowledge and routines to solve problems (Star, 2005). Procedural fluency can be supported instructionally by modeling specific strategies to solve mathematical practice, providing adequate opportunities for practice, and giving students feedback (Miller and Hudson, 2007; NCTM, 2000, 2006; National Research Council, 2001). Second, high-quality instruction attends to *conceptual understanding* and the development of mathematical concepts. Conceptual understanding is central to mathematical proficiency (National Research Council, 2001), and it is particularly important in providing students with instructional opportunities that develop their conceptual understanding of the big mathematical ideas within the kindergarten through 12th-grade (K–12) curriculum (Bransford, Brown, and Cocking, 1999; NCTM, 2000; National Research Council, 2001). Scholars have defined conceptual understanding as a function of the strength of the relationships between ideas (Hiebert, 2013; Goldman and Hasselbring, 1997; NCTM, 2000). Building conceptual understanding facilitates the development of a knowledge network in which ideas are strongly linked to one another. High-quality instruction encourages students to develop this knowledge network by making connections to prior learning and big ideas across grades and subjects. This includes connections between previously learned mathematics (Bulgren et al., 1995; Miller and Hudson, 2007; NCTM, 2000, 2006), as well as opportunities for students to apply existing knowledge to novel contexts (Goldman and Hasselbring, 1997; Miller and Hudson, 2007; NCTM, 2000, 2006).

Conceptual understanding can be supported instructionally by using problem-based instruction (Lampert, 1990; Hiebert and Wearne, 1993; Lappan and Phillips, 1998; NCTM, 2000, 2006; National Research Council, 2001; Stein, Boaler, and Silver, 2003) that incorporates manipulatives or other tools that facilitate exploration of mathematical ideas (Shaw, 2002; Miller and Hudson, 2007).

It is also critical that students be provided opportunities to engage in cognitively demanding mathematical tasks (Stein, Grover, and Henningsen, 1996; Doyle, 1983; Charalambous, 2010; Boston and Smith, 2009) in order to facilitate conceptual understanding (Charalambous, 2010; Boaler, 2002; Boaler and Staples, 2008). Cognitively demanding tasks are distinguished by several features. They can be solved and represented in multiple ways, including symbolically and with manipulatives. Conceptual understanding is enriched by making connections among these various representations (Boston and Smith, 2009). Cognitively demanding tasks encourage students to make sense of what is being asked by requiring complex and nonalgorithmic thinking (Boston and Smith, 2009). This is often achieved by encouraging students to apply mathematics to model real-world phenomena. Cognitively demanding tasks also encourage students to explain their own thinking, make viable arguments, and critique the reasoning of others (Common Core State Standards Initiative, undated [b]; NCTM, 2000, 2006).

Equity and Diversity

Access to a common curriculum has long been central to equalizing educational opportunities; while this goal has been elusive in the past (Coleman, 1968; Murphy, 1988), policymakers and U.S. courts have strived to make this a reality. Currently, the goal of a successful CCSS implementation supports the achievement of a curriculum-based focus on education equity and consistent expectations across the nation (Equity Assistance Centers, 2013; Common Core State Standards Initiative, undated [a]). Rigorous standards alone, however, do not address the issue of the achievement gap, which is defined as the difference in academic performance between groups of students (U.S. Department of Education, undated). Education equity in the context of CCSS implementation, therefore, requires a standard way to measure all student groups on achievement indicators to track performance and supports for all students in order to facilitate their learning (Equity Assistance Centers, 2013). It is clear that comparable positive outcomes for all students can be achieved only through monitoring the progress of all students in the classrooms, schools, districts, and states. This can be measured with disaggregated test scores, attendance data, promotion and graduation rates, and all other relevant outcomes (Equity Assistance Centers, 2013).

Maintaining the inclusion of all students in the CCSS curriculum is vital to reaching the goal of education equity, but this can be achieved only through equitable access to education services for all students that, at the same time, recognizes the need for differentiated instruction when necessary and provides available systems and interventions to meet individual students' needs (Hakuta and Santos, 2012, p. 2; Equity Assistance Centers, 2013; Moschkovich, 2012; Walqui and Heritage, 2012; Powell, Fuchs, and Fuchs, 2013; Common Core State Standards Initiative, undated [a]; O'Day and Smith, 1993).

For example, although the Individuals with Disabilities Education Act (Pub. L. 101–476, 1990) requires that students with special needs be educated in the "least restrictive environment" and that high-needs populations be taught alongside all other students, O'Day and Smith, 1993, argue that schools may require different instructional, curriculum, or personnel resources to educate all students well. Similarly, Hakuta and Santos, 2012, suggest that ELs "have a right to appropriate education that is grounded in sound theory and implemented in ways that address their needs systematically, through coordinated support linking teachers, materials, formative assessments, tests and accountability systems, and technology" (Hakuta and Santos, 2012, p. ii). Moreover, because the CCSS do not define advanced work and post–high school standards, it can fall short in meeting the specific needs of advanced and gifted learners (Common Core State Standards Initiative, undated [a]). Schools may need to create additional supports for advanced learners through differentiated curriculum and instruction.

Teacher Content Knowledge

There is considerable research demonstrating that effective teaching requires rich mathematical understanding (NCTM, 2000; Wu, 2011). Much of this research focuses on what has been called Mathematical Knowledge for Teaching (MKT), a concept that was developed by Ball and colleagues to describe the specialized content knowledge needed for teaching mathematics (Ball, Hill, and Bass, 2005; Ball, Thames, and Phelps, 2008; Hill, Ball, and Schilling, 2008). MKT consists of

- common content knowledge (i.e., knowledge of commonly used mathematical concepts and procedures)
- specialized content knowledge (i.e., knowledge of why procedures work and knowledge of which particular representations are most helpful for student sense-making)
- knowledge of content and students (which exists at the intersection of knowledge of mathematical content and knowledge of how students learn)
- knowledge of teaching (which exists at the intersection of content and pedagogy).

Studies have shown that higher MKT is positively correlated with a teacher's ability to select and enact mathematically rich tasks in elementary school classrooms (Charalambous, 2010), higher quality of instructional practices (Hill, Kapitula, and Umland, 2011), and student achievement in mathematics (Hill, Rowan, and Ball, 2005).

Classroom Assessment

Assessment is a critical feature of teaching and learning. High-quality, thoughtful, and thorough assessment programs contain not only *summative* assessments, which are intended to measure progress or learning, but also *formative* assessments, which can provide critical insights into how students think and allow teachers to adjust instruction accordingly (Darling-Hammond et al., 2013; Black and Wiliam, 1998; Brookhart, 2011).

Formative assessment is sometimes referred to as assessment *for* learning (Brookhart, 2011) because it is concerned with using information about the quality of student responses to improve student competence (Sadler, 1989). Because high-quality teaching generally involves an iterative process of teaching, checking for understanding, and reteaching as necessary, some researchers have noted that high-quality formative assessment is deeply embedded in high-quality instruction (Wiggins, 1998). An integral part of formative assessment involves using student work to inform next steps in teaching and learning, and accomplished mathematics teachers use a variety of ongoing assessment strategies (both formal and informal) to diagnose learning and plan instruction (Hattie and Timperley, 2007; Brookhart, 2011). Formative assessments can include assessments of content knowledge, procedural fluency, and students' perceptions of mathematics as a discipline (McIntosh, 1997). Strategies for formative assessment can include daily quizzes; student interviews; the use of student self-rating scales; daily exit cards (written student responses to questions teachers pose at the end of a class or lesson); and cognitively demanding performance tasks, such as those described above (McIntosh, 1997).

A summative assessment is defined as a "culminating assessment, which gives information on a student's mastery of content" (Association for Supervision and Curriculum Development [ASCD], 1996, p. 60). In the context of standards-based education, it is critical to consider the alignment between summative assessments and the standards they purport to measure (Darling-Hammond et al., 2013; Porter et al., 2011). In other words, assessment items should closely reflect the content and processes that are articulated in standards. In fact, the alignment of standards and assessments was legislatively mandated by No Child Left Behind to satisfy its rigorous requirements for accountability (U.S. Department of Education, undated[b]; Polikoff, Porter, and Smithson, 2011) and is a critical component for establishing the validity of inferences about student mastery that can be made based on assessment performance (Kane, 2008).

The CCSS in mathematics include both the Standards for Mathematical Content, which describe the knowledge and procedures that students should master, and the Standards for

Mathematical Practice, which describe the mathematical behaviors and dispositions—such as perseverance, abstract reasoning, modeling, argumentation, and reasoning—that students should possess. The literature notes that a high-quality and valid summative assessment program should give a balanced consideration to the assessment of the Standards for Mathematical Content and the Standards for Mathematical Practice (Kepner and Huinker, 2012; Krupa, 2011), meaning that assessment should contain a balance of tasks that appraise students' content knowledge and the behaviors and types of expertise that constitute desired mathematical practices (Kepner and Huinker, 2012). Researchers have also warned that teachers often focus on the material they know will be included in summative assessments (Wilson, 2007; Krupa, 2011). Consequently, excluding the Standards for Mathematical Practice from summative assessments compromises the successful implementation of the CCSS (Krupa, 2011).

Beyond adequate alignment, researchers have recommended that summative assessments be supported by technology and have multiple modes of content delivery (Krupa, 2011), as traditional paper-and-pencil assessments will not be adequate to measure student mastery of the CCSS (Lazer et al., 2010). Additionally, digital resources allow for a much wider range of assessment formats and more-sophisticated item types. As such, technology permits flexible and adaptive testing, which can reduce the assessment burden currently placed on students by fixed-length traditional assessment (Lazer et al., 2010). Another benefit that researchers have noted is that the use of computer and online assessments can reduce the amount of time that educators must wait for achievement data (Tamayo, 2010).

Instructional Leadership

There is an abundance of literature on effective leadership in general and on how school leaders can prepare their schools and teachers for shifts embodied in new standards, such as the CCSS or the CCR standards (Achieve et al., 2013; Education First and Achieve, 2012; Aspen Institute Education and Society Program et al., 2013). A widely used framework from the general literature on effective school leadership developed in Bossert et al., 1982—and more recently updated by Leithwood et al., 2004—identifies three common practices that make up successful leadership: (1) setting direction, (2) developing people, and (3) ensuring that organization design facilitates the achievement of school vision. Recent research provides further empirical support for this framework (e.g., Grissom, Loeb, and Master, 2013).

One of the main components of leadership's direction setting is identifying and articulating a vision (Leithwood et al., 2004; Riddle, 2012; National Association of Elementary School Principals [NAESP], 2012). Generally, this vision should be defined for curriculum, instruction, assessment, and intervention (Kanold and Larson, 2012; ASCD, 2012). The vision is a unifying and focusing element that can increase teacher engagement and help to build staff trust (Achieve et al., 2013; NAESP, 2012). The principal should also help subunits, such as teacher teams, focus their work according to the vision and aid unit-by-unit planning. It is also important to communicate the vision with families and community stakeholders (ASCD, 2012). Within the context of the CCR standards, the principal would need to clearly communicate the vision of the new standards adopted, how the vision differs from the old standards so all parties involved (e.g., teachers, parents, students) understand the changes students will be facing, the goals of the curriculum or standards, and what educational stakeholders' roles are both inside and outside of the classroom (Leithwood et al., 2004; Larson, 2011).

To aid in achieving a new school vision, such as the adoption of new standards, the school leader should set clear and high expectations for the academic, social, emotional, and physical development of all students (Achieve et al., 2013). In order to verify that these expectations are being met, the school leader will need to develop or alter means for measuring and assessing progress (Achieve et al., 2013). The means may include the use of data to monitor student, teacher, and school performance; observations that measure students' engagement in instruction; and tracking teacher progress toward implementing the new vision for standards, curriculum, and instruction, including collecting teacher feedback (Kanold and Larson, 2012; Achieve et al., 2013; Leithwood et al., 2004).

It is necessary for leadership to build capacity among teachers and staff in order to work toward the goals and expectations determined in direction setting. The school leader is responsible for providing an effective approach to PD that includes learning opportunities for the staff, fosters intellectual stimulation, and offers individualized support (Leithwood et al., 2004). Within the context of the CCR standards, a commonly recommended model is to establish a leadership team for implementation comprising teacher and administrator leaders from diverse groups (Achieve et al., 2013; Riddle, 2012; ASCD, 2012; Leithwood et al., 2004). This leadership team is responsible for taking charge of CCSS knowledge, aiding in the development of an implementation plan, and helping to identify areas for growth. This group of experts in the building can also facilitate the induction of new teachers, as well as mentoring and coaching of all teachers. In many cases, the aforementioned leadership team can assist in targeting the areas that need the most attention (Leithwood et al., 2004). The principal can then set short- and long-term plans for continuous, connected, ongoing, and job-embedded PD. While the focus of PD may vary depending on school needs, most of the literature recommends incorporating PD that focuses on deeper knowledge of standards, curriculum changes, and instructional practices (Achieve et al., 2013; NAESP, 2012; Larson, 2011; Education First and Achieve, 2012).

Ensuring alignment between organizational cultures and structures and school vision can also facilitate progress toward school goals. Major school initiatives may require changing school cultures, modifying organizational structures, or building new collaborative processes (Leithwood et al., 2004). With regard to the implementation of the CCSS, Achieve et al., 2013, recommends creating a culture of continuous learning and collaboration among teachers that is tied to student learning and other school goals. Sufficient time should be provided for teacher collaboration in professional learning communities, which are best organized by grade level where possible (Larson, 2011; Kanold and Larson, 2012). School leaders may also adapt hiring and induction processes to ensure that new staff deliver standards-aligned instruction with fidelity (Riddle, 2012; Leithwood et al., 2004). For mathematics, in particular, principals could also make assignment decisions based on CCSS-related qualifications and demonstrations of teacher effectiveness in teaching math (Larson, 2011).

Administrator and Teacher Professional Development

For many teachers, successful implementation of the CCSS will require changes to their instructional practice (Wu, 2011; Ball and Forzani, 2011). However, research has shown that teachers are resistant to change (Goldenberg and Gallimore, 1991) and that it is difficult to deliver effective PD that changes instructional practices (TNTP, 2015). Nevertheless, the literature highlights the characteristics of promising PD. For example, Correnti and Rowan, 2007, compares three different instructional programs and reports the authors' finding that only programs that offered strong support by on-site facilitators and local leaders facilitated changes in instructional practice. Other research found that teachers were more likely to change their instructional practice when participating in PD programs that offered ongoing on-site support and teacher networks that could work collaboratively within and across schools and districts (Corcoran, McVay, and Riordan, 2003). This is consistent with a larger body of scholarly work suggesting that effective PD is sustained and intensive (Marrongelle, Sztajn, and Smith, 2013; Loveless, 2013) and involves collective participation and group work (Desimone, 2009; Elmore, 2002; Guskey and Yoon, 2009; Wei et al., 2009; Elliott et al., 2009).

There is some consensus that PD must involve active learning in order to be effective (Correnti, 2007; Correnti and Rowan, 2007; Garet et al., 2001; Supovitz and Turner, 2000). Active learning (Garet et al., 2001) can include such activities as opportunities to observe master teachers, curriculum mapping, and lesson planning. Active learning can also include opportunities to engage directly in the analysis of student work (Darling-Hammond, 1997) and opportunities to engage in mathematical problem-solving (Borko, 2004).

Additionally, effective PD is integrated into daily activities and is focused on specific curricular content (Correnti, 2007; Correnti and Rowan, 2007; Garet et al., 2001; Supovitz and Turner, 2000). A focus on content is critical because the CCSS require more-sophisticated mathematical content knowledge to support student work and to help students develop mathematical proficiency (Wu, 2011; Ball and Forzani, 2011). Studies of cognitively guided instruction (Carpenter et al., 1989), the problem-centered mathematics project (Cobb et al., 1991), and the educational leaders in mathematics project (Simon and Schifter, 1991) all demonstrated that PD programs that were content focused could have positive impacts on instructional practice and student achievement by improving teachers' understanding of mathematics and student mathematical thinking.

CCSS authors stress that PD should not only focus on building teachers' capacity to implement the content standards but also include careful attention to the practice standards ("Gearing Up for the Common Core State Standards in Mathematics: Five Initial Domains for Professional Development in Grades K–8," 2011). Additionally, because the CCSS require more-sophisticated knowledge and many teachers report being unprepared to implement the standards successfully (Walters et al., 2014), several advocacy groups (such as Student Achievement Partners) have recommended that initial PD programs focus on building an understanding of the features of the CCSS that are different from those of existing standards frameworks. In mathematics, these "shifts" entail (1) greater focus on fewer topics, (2) linking topics and thinking across grades, and (3) rigorous pursuit of conceptual understanding (e.g., Alberti, 2013).

Features of effective PD for school leaders largely overlap with the features of high-quality teacher PD previously cited, including (1) integration into daily activities, (2) ongoing and sustained participation, and (3) opportunities for collaboration (Goldring, Preston, and Huff, 2012; Lawrence et al., 2008; Evans and Mohr, 1999; National Staff Development Council, 2000). School leaders also have some responsibilities that are specific to their role, and these responsibilities create unique PD needs. Several scholars point to the importance of providing PD opportunities that are differentiated based on leadership experience (Kelley and Peterson, 2002; Peterson, 2002; Goldring, Preston, and Huff, 2012). Additionally, Goldring, Preston, and Huff, 2012, notes that, unlike teachers, school leaders are often isolated in their buildings. Because of this, opportunities for networking and professional support are particularly important for reflecting on and improving practice.

Access to STEM Enrichment

STEM skills are becoming increasingly essential for participation in the globalized economy. School settings provide time for STEM learning, but offering additional STEM opportunities outside the classroom through expanded and extended programs is viewed by many educators as important to increase student access to and engagement in STEM (Barron, Wise, and Martin, 2012; Khisty and Willey, 2012; Krishnamurthi, Ottinger, and Topol, undated; Cullum et al., 2008; Hynes and Dos Santos, 2007). Research suggests that STEM learning opportunities can be enhanced through integrative approaches within classrooms, extended programs, or expanded programs (Sanders, 2009, p. 21; Bevan and Michalchik, 2013). Integrative approaches are defined as "approaches that explore teaching and learning between/among any two or more of the STEM subject areas, or between a STEM subject and one or more other school subjects" (Sanders, 2009, p. 21). Extended and expanded programs should both be considered as viable additions to a regular curriculum; however, both would be provided after regular school hours. Extended programs align more closely with the school curriculum, while expanded programs address subject matter and practices that are not included in the grade-level curriculum (Bevan and Michalchik, 2013).

Regardless of which STEM approach is adopted, providing consistent and ongoing programming that extends across grade levels and making it available to all students can increase student interest in STEM and improve STEM-related skills (Krishnamurthi, Ottinger, and Topol, undated; Bevan and Michalchik, 2013). STEM learning programs that focus on offering hands-on experiences not only encourage children to become engaged in these topics, but also help them to build real-life skills (Cole, 2011; Krishnamurthi, Ottinger, and Topol, undated; Afterschool Alliance, 2013). Partnerships with STEM professionals and STEM-rich institutions can also support the development of quality and robust STEM learning programs (Eccles, 1994; Halpern et al., 2007; Liston, Peterson, and Ragan, 2007; Koch et al., 2010; Afterschool Alliance, 2013). Having STEM professionals serve as role models can have positive influences on students, enhance their perceptions of STEM careers, and boost their confidence in studying such subjects. This is more likely to be the case if STEM professionals engage directly with students, demonstrating what they do and connecting their skills to real-life situations, rather than simply giving overviews of their jobs in conversations with students (Cole, 2011; Afterschool Alliance, 2013). The possibilities for hands-on opportunities with STEM professionals and STEM-rich institutions can, of course, vary greatly. They can range from going to science centers, where students have access to tools and instruments intended for hands-on learning, to helping university graduate students with research experiments.

In addition to the literature summary presented, we interviewed three LEAs about their experiences in adopting and implementing the CCSS in mathematics to inform the best practices we delineated. The section below presents findings from the LEA interviews.

Local Education Agencies' Lessons Learned

We contacted exemplary LEAs to obtain a richer, more-specific understanding of their practices in the context of the CCSS in mathematics. The research points to the importance of offering PD that involves teachers in active learning, encourages collective participation, and focuses on content of standards. The LEAs involved teachers and coaches along with district staff early in the process to prepare them for the new standards. Specifically, at the elementary, middle, and high-school levels, teacher work groups were formed. Some LEAs involved all their teachers in the work groups, while others involved representatives from each school or grade level. The work groups met during the summer prior to the first year of implementation. They were asked to review the mathematics program they were implementing and the new standards, to identify which aspects of the program were aligned with the new standards, and to eliminate topics that were not grade or standard relevant. The work groups also researched teaching techniques. One of the goals of the work groups was for teachers themselves to narrow the standards they would emphasize and explore them in greater depth. LEAs indicated that this approach served as an intensive PD and set expectations for future practice. The work groups reviewed and discussed the new standards and, in the process, gained in-depth understanding of the standards' content and the practices required to address the standards in the classroom. Further, this approach increased teacher buy-in and promoted a sense of ownership of the reform early on.

The literature also highlights the importance of providing teachers with curriculum resources to support standards-aligned instruction, and the LEAs emphasized this point. The LEAs recognized the need for making such resources available, and they engaged teachers in creating them. Teachers who were identified as leaders or who volunteered to be involved worked with coaches to develop curriculum aligned with the standards. The teachers and the coaches took existing pacing guides, aligned them to the new standards, and communicated to the rest of the teachers the kinds of changes they needed to make, without dictating specifically how each unit or lesson should be taught.

The LEAs also emphasized the need to provide ongoing support to teachers to facilitate implementation over time, but they noted the challenges of providing such support, particularly the challenges associated with time and resources. The support provided by the LEAs varied and included embedded professional learning days, online videos and courses on the application of standards for mathematical practices in real settings, and opportunities for teachers to observe "exemplary classroom instruction" and engage in classroom walk-throughs. Embedded PD included teachers getting together to analyze student data and then collaboratively design lessons based on data they analyzed, followed by having one teacher model the lesson in a classroom while others observed and later discussed what they observed. One of the LEAs indicated the need to build professional learning communities among teachers to sustain continued teacher development.

Indicators of Best Practices

In evaluating DoDEA's mathematics program, we derive best practices from the literature review and LEA interviews. Table 2.2 presents the practice indicators for each of the eight programmatic areas under study. These practices are used to evaluate the quality of DoDEA's mathematics program.

Table 2.2
Best-Practice Indicators

Mathematics Program Area	Indicator
Curriculum resources	• Align with mathematics standards • Support teacher standards–aligned instruction
Curriculum and instructional quality	• Focus on fewer topics but in greater depth • Link major mathematics topics within and across grades • Pursue conceptual understanding and procedural skills with equal intensity • Engage students in cognitively demanding tasks • Model with mathematics to solve problems in the real world • Use appropriate tools and manipulative aids • Encourage students to explain, make viable arguments, and critique reasoning of others
Equity and diversity	• Monitor the performance of all student groups on relevant outcomes • Have in place systems and interventions to meet needs of all students • Ensure equitable access to curriculum and promote student inclusion • Differentiate instruction during class to meet differing student needs
Teacher content knowledge	• Ensure that teachers have an in-depth understanding of mathematics and the mathematics standards
Classroom assessment	• Offer a thoughtful and thorough formative assessment program • Align with standards • Place equal emphasis on content and practices • Support with technology
Instructional leadership	• Have leaders set direction • Have leaders develop people • Have leaders ensure that organization design and vision are aligned
Administrator and teacher PD	• Be sustained and intensive • Be content focused • Involve active learning • Integrate into daily activities, with collective participation • Focus on CCR standards "shifts"
STEM opportunities	• Incorporate STEM professionals as role models • Partner with STEM-rich institutions • Have partners demonstrate and do, rather than talk and listen • Offer regular, consistent programming • Include cross-grade activities

CHAPTER THREE

Evaluation of the Current DoDEA Mathematics Program

In this chapter, we describe the practices implemented by DoDEA schools and teachers in each of the eight key areas identified by DoDEA and discuss variations across sites where those were noteworthy. For each of the areas, we also examine the extent to which the schools' practices are aligned to the best practices highlighted in Chapter Two.

Curriculum Resources

We identified two key indicators of best practices with respect to curriculum resources:

- alignment with mathematics standards
- support for standards-aligned instruction.

For teachers to implement standards-based instruction, high-quality curriculum and resources aligned with the standards need to be available. In addition, teachers should be provided with enough guidance on how to implement the materials to ensure that the resources are used in an optimal way to support student learning.

At the time of our visits, DoDEA schools were implementing the mathematics standards adopted in 2009. Schools were using mathematics textbooks that DoDEA recommended with the 2009 standards. The textbooks included *Everyday Mathematics* for grades K–2, *enVision* for grades 3–5, and *Mathematics Connects* for middle school grades. A variety of textbooks was used for different courses (e.g., algebra, geometry) in high schools. The extent to which teachers relied on those textbooks to teach the standards varied among teachers, grade levels, and sites. In interviews, some teachers who reported the textbooks to be aligned with the mathematics standards relied on the books, although they skipped chapters or units that did not address the standards. Many teachers, however, indicated that the textbooks were not well aligned with the mathematics standards. They talked about using other resources either as supplements or as main curriculum materials. As one high school teacher put it,

> We supplement all the time. I think, with mathematics, you need more than the textbook. Sometimes mathematics textbooks are difficult to read and understand so that you supplement by either PowerPoint or videos or someone going through examples with them. In my class, I look for activities to do with students. Again, it is that balance. Can I find something that can teach them [a particular topic] and engage them? I use the textbook as a reference.

Different reasons were given for seeking outside resources. In most interviews, teachers indicated that they sought additional resources to augment classroom and homework exercises and assessments included in the textbooks. IXL, an online mathematics software program, was frequently used with elementary and middle school students to practice mathematics skills. A number of elementary schools whose continuous school improvement (CSI) goal focused on problem-solving used Exemplar (a publisher of hands-on, standards-based assessment and instructional material that focuses on authentic learning [an approach that encourage students to learn through hands-on, collaborative projects that address problems relevant to their lives] in the areas of math, science, and writing) supplemental materials. In these schools, principals and teachers emphasized the CSI mathematics goal and reported integrating Exemplar into their curriculum because it provides teachers and administrators with a way of assessing students' problem-solving and communication skills using real-world scenarios. Interviews also revealed other, less frequently raised reasons for relying on additional resources. Some elementary school teachers reported that the textbooks provided by DoDEA either lacked emphasis on developing students' mathematical concepts or did not include materials and activities that teachers could use to support struggling students and reteach them concepts they had not mastered. Some middle school teachers incorporated additional resources, such as Khan Academy videos, to help students who were not performing at grade level. At the high-school level, teachers tended to use a compilation of external resources for instruction. This is especially the case in flipped classrooms, in which the video lecture that students watched on their own time was seen as the key ingredient rather than the textbooks. In Europe, a number of DoDEA teachers created video lectures and shared them with their colleagues for use. Other videos were selected from an online repository.

Interviews indicated that, overall, there was no defined process throughout the DoDEA system to identify resources and examine their quality. Individual teachers tended to seek their own resources when the need arose and then shared them with their peers. In a couple of districts, the mathematics ISS reported creating a scope-and-sequence document that aligned the concepts to be covered in the standards with the DoDEA textbooks and identified additional relevant resources and activities. Some of the teachers interviewed in that district expressed the value of having a scope-and-sequence document to link standards with a sequence of textbook-based lessons and supplemental materials. One teacher noted,

> The most basic structure is that [the ISS] sends out scope and sequence and collaborative activities that you can use throughout the course. [The ISS] provides nice collaborative constructivist lessons and the support to do them. This is very helpful for teachers.

Some teachers also noted that the lack of a unified curriculum scope and sequence across DoDEA districts has negative effects on mathematics education for military-connected students who move between schools frequently.

Curriculum and Instructional Quality

Our literature review identified three indicators of a high-quality, standards-based curriculum:

- *focus* on fewer topics but in greater depth

- *coherence* among mathematics topics across grades and subject areas
- *rigor* of mathematical thinking, placing equal emphasis on conceptual understanding and procedural fluency.

Further, the literature review identified four features of high-quality instruction:

- Engage students in cognitively demanding tasks.
- Model with mathematics to solve problems in the real world.
- Use appropriate tools and manipulative aids.
- Encourage students to explain, make viable arguments, and critique the reasoning of others.

Principals and district ISSs described having too many topics addressed in their 2009 standards-based curricula, reporting that the number of standards has made it challenging for teachers to implement the mathematics program. One ISS echoed this view and said,

> DoDEA standards, in general, have been difficult to implement. This is because there are so many of them. You usually want to pick two or three. In DoDEA, you have 12 standards for Number Sense. What do you focus on; which one is more important? So many standards to pick out, and you have students with different abilities in class and you want to differentiate. It is too much.

A few principals noted that it is difficult for teachers to help their students master the standards when there are so many. They indicated that their teachers reference the standards but do not teach all in depth. Some teachers also reported that they focus on teaching those standards addressed in the TerraNova or Preliminary SAT/National Merit Scholarship Qualifying Test (PSAT/NMSQT) as a way to limit their number and ensure they are taught in-depth.

The literature points to specific practices associated with effective mathematics programs in addition to mastering content. Teachers of high-quality programs help their students make connections between mathematics concepts. Of the teachers surveyed, 91 percent reported placing "moderate to high" emphasis on making connections between mathematics concepts at their grade level, while 81 percent indicated emphasizing mathematical connections across grade levels at least moderately. However, in only 50 percent of the classrooms observed did teachers make explicit connections between key mathematics concepts. Mostly, teachers referred to mathematical concepts or procedures they covered earlier in the week with no discussion or deep exploration of those connections by either the teacher or the students. For example, in one geometry class, the teacher showed students that a trapezoid could be divided into two triangles and referred to the triangle area formula taught earlier in the week to help students calculate the area of the trapezoid. In another class, we observed the teacher referring to the need for understanding the distributive property before moving onto algebraic expressions without engaging students in a discussion about this relationship. We did not observe many teachers making connections from the context of the problem to a mathematical representation, between different mathematical representations, to previous learning, or to future work. In the elementary grades, different textbook series across grades present an additional obstacle to making connections between current mathematics content and prior or upcoming content. The Everyday Mathematics curriculum used in grades pre-K–3 uses different vocabulary and approaches from the enVision curriculum used in grades 4–6. It may be that teachers emphasize connections between mathematical concepts (as they reported on the survey) but without making these connections in every lesson.

Making connections across disciplines was less frequent. Teachers we observed connected lesson content explicitly to other disciplines in 20 percent of the classes. For example, in one of the lessons that included coin-counting activities, the teacher referred to the U.S. presidents who are on the currency and mentioned to the children some historical facts. We also observed teachers reinforcing the English language to their ELs by having them read a mathematics word problem and articulating their understanding of the problem. Artifacts collected from approximately 80 percent of the observed lessons did not show any explicit connections between mathematics topics. Few lesson plans connected mathematics activities together or delineated activities in which students have to recall what they learned in the previous lesson. Classroom and homework assignments comprised mostly "drill exercises."

Another best practice identified in the literature is incorporating classroom activities that develop students' conceptual understanding, procedural skills, and application. Approximately 90 percent of teachers surveyed indicated they put "moderate to major" emphasis on having students practice computation and solving unfamiliar problems that require mathematical thinking. Our observations showed that teachers had difficulty balancing their activities to promote students' mathematics understanding and fluency in procedures and skills. In 60 percent of the observed classrooms, teachers spent the majority of their time covering concepts through lecturing, modeling mathematical procedures for concepts taught, and asking students to apply the same procedures to solve a set of problems similar to those solved by teachers. We did observe teachers in about one-third of the classrooms successfully implement challenging activities that stretched beyond just procedural skills. Teachers had students set up problems, engage in novel modeling or postulate testing, and determine the correctness of the answers. For example, we saw a group activity in a high school geometry class in which students were asked to prove various postulates regarding congruent triangles, including "side-side-side," "angle-side-angle," and "side-angle-side." The teacher asked students to try to create incongruent triangles that were still consistent with these postulates. Students worked in groups attempting to create incongruent triangles using specific lengths and angle measurements provided by the teacher. The teacher checked student work throughout the activity. For those groups that managed to create what appeared to be incongruent triangles that were consistent with the measurements and the postulates even though this was impossible, the teacher did not just correct their work but instead asked them to rethink their efforts and start over. In another example, one teacher gave her middle school students problems, such as the following and asked them to set up a system of equations to help solve the problem:

> Nancy's farm has ten animals. Her animals have a total number of 32 legs. On her farm she has cows with four legs each and chickens with two legs each. How many cows and chickens does she have?

This activity was challenging because the students were required to set up the system of equations on their own. Students needed to put various pieces of information together to build the equations. At each step of the task, students needed to make decisions and implement a strategy to find the right answer. Students in the class struggled with the activity, but the teacher provided them with adequate time to discuss it with their partners. Our analysis of the artifacts indicated that the majority of lesson plans and homework assignments did not draw out new ideas or challenges and mostly focused on accuracy, skill acquisition, and procedures.

Mathematical modeling is an important tool that can be used for problem-solving and forecasting, and it reinforces the application of mathematics concepts in real-life situations (Council of Chief State School Officers [CCSSO], 2013). Moreover, as conceived of in the CCSS, mathematical modeling also provides opportunities for students to practice communication about mathematics, because one of its key components is having students work in pairs or groups to develop, explore, and test the appropriateness of models (CCSSO, 2013). Modeling is one way to link classroom mathematics and statistics to everyday life, work, and decisionmaking (CCSSO, 2013).

Of the teachers surveyed, 89 percent reported placing "moderate or major" emphasis on having students in their classes apply mathematical principles in real-life situations. In our observations, about 28 percent of teachers provided their students with opportunities to solve authentic and realistic problems arising in everyday life. At the elementary level, modeling activities included modeling a rocket launch to observe, measure, and work with Newton's third law and using simple arithmetic to solve word problems addressing real-life situations. At the higher grades, students were given opportunities to model relationships between two variables. For example, in one of the classes, the teacher presented students with a realistic situation of opening a new business that makes and sells vases. Students were asked to determine the point at which the business will break even. This scenario required students to make various assumptions and identify the y-intercept of two equations representing the cost of making vases and the income generated by selling vases. In terms of artifacts, very few, representing 18 percent of the observed lessons, included mathematics problems with real-world aspects that required students to demonstrate their understanding.

Tools and manipulatives were available in most of the classrooms we visited. Of the teachers we observed, 38 percent used tools and manipulatives to develop students' conceptual skills. Teachers, for example, made tools available for students to measure body dimensions. Calculators were provided to students in some classrooms. Place-value blocks were also used in some classes to assist students in creating mathematical models of real-life relationships. Even though technology (e.g,. calculators, TI-Nspire™, Smartboards®) were widespread, they were used mostly to reinforce procedures. Furthermore, Smartboards were being used as traditional white boards to present information or video to the whole class or student groups. Very few teachers incorporated the interactive features of the Smartboards into their lessons to promote student interaction.

We also observed very few teachers giving students greater responsibility for developing mathematical concepts while the teacher acts more as a learning facilitator. In most classrooms, teachers initiated and led most discussions of mathematical concepts. Over half of the observed teachers asked procedural questions or questions requiring yes/no answers. About 46 percent used questioning strategies to encourage students to clarify their thinking and show their understanding, but the majority of these interactions did not foster discussions among students. Furthermore, most teacher questions did not encourage students to clarify and extend their thinking, probe deeper, or make connections. Very few teachers asked questions that focused on promoting mathematical reflection (e.g., Why does your answer make sense?) or a deep understanding of mathematical ideas (e.g., Is this true in all cases?).

Equity and Diversity

The literature stresses the importance of implementing inclusion policies to ensure that all students have equitable access to the curriculum and education services. But the literature also adds that inclusion policies are insufficient on their own and that it is necessary for such policies to address differentiation in instruction to take into account varying student abilities and learning styles in a classroom. To achieve this goal, it is important for schools to put in place a system that measures all student groups on achievement indicators and uses the information to track each group's performance, identifying different student needs and providing the appropriate support to each student.

The districts and schools visited engaged in efforts to review student assessment data to varying degrees. All principals and teachers interviewed reported that they review student assessment information to monitor student groups and identify which mathematics areas to emphasize. The type of assessments schools used and the purpose of the assessment review differed across schools and by grade level. At the elementary and middle schools, principals and teachers emphasized reviewing school assessment data (e.g., schoolwide standardized assessment in selected schools or Exemplar data) and formative in-class assessments and, to a much lesser extent, stressed studying the results from the systemwide TerraNova assessments. TerraNova was not considered by either principals or teachers as a valid tool for monitoring student performance. One elementary school principal said, "TerraNova is one source of data, but I do not like it. It is not diagnostic and, by the time I receive the data, it is late." Some elementary school principals reported reviewing data from their school-specific assessments at the end of each quarter during their leadership meetings to determine student support. Other principals administered pre- and postquarterly assessments to set school goals and monitor student progress toward the goals. Students not meeting the school goals were regrouped for instruction. A few middle school principals reported assessing incoming fifth graders at the beginning of the school year to determine their course placement. At the high-school level, ninth-grade TerraNova results as well as PSAT/NMSQT information were commonly used to identify the mathematics areas on which students needed improvement. For the TerraNova, some high school principals reported reviewing basic statistics (e.g., median, mode, and mean) for various student groups, as well students' objective performance index, to determine mastery levels in mathematics. Others preferred to review PSAT/NMSQT results because they could generate a list of questions that were answered incorrectly to inform mathematical areas on which to focus. Some principals also looked at course grades and placed students earning Ds or Fs on a list for weekly monitoring. District ISSs' efforts in using assessment data to address achievement gaps were limited because they had access to TerraNova results and classroom observation data only from the classrooms they visited. Finally, some teachers in high schools used departmental assessments to identify low-performing students for ongoing monitoring. Although principals and teachers frequently used data to identify and monitor students who were performing poorly, it was unclear whether they monitored with the same diligence student performance by gender or ethnicity.

Interviews at all school levels noted the lack of available mathematics interventions for low-achieving students, such as resources that provide opportunities for students to work with visual presentations of mathematical ideas or computer-adaptive mathematics software that regularly assesses students' mathematics knowledge, tracks student growth, and helps guide instruction or other intensive mathematics interventions designed to improve the skills of low-achieving students.

Most of the supports that were mentioned by principals and teachers irrespective of the school level were in the form of extra time spent with low-performing students, such as tutoring, homework club, seminars, or "opportunity halls." Few schools with local mathematics-support specialists mentioned pulling out low-achieving students to provide them with additional instruction. The mathematics-support specialists also provided in-class support upon teacher request. Low-achieving students who did not improve as a result of the supports were referred by teachers to the student support team (SST). Both teachers and SST members work together to analyze student misunderstandings and design interventions to effect the desired change.

The literature emphasizes the importance of expecting ELs and students with special needs (special education students) to achieve CCR standards and to ensure that those student groups have the same access to the curriculum as their peers. The literature also points to the need to provide appropriate instruction for gifted students, and it is the school's responsibility to create programs or courses to support the educational needs of advanced learners. Most identification of EL, special education, and gifted students occurs at the elementary school level. Their identification status transfers with them as they move through higher grade levels. Some identification occurs in later grades, particularly for students who enroll in DoDEA schools for the first time in the secondary grades.

English Learners and Students with Special Needs
DoDEA schools have effective and consistent systems in place for identifying students who are not proficient in ELA and students with other learning needs.

To identify ELs, all new families are asked to complete the Home Language Questionnaire at the time of registration, regardless of the student's language, race, or ethnicity. Questionnaire information is provided to the English-as-a-second-language (ESL) teacher at the school whenever the parent has indicated that a language other than English is spoken in the home. The ESL teacher reviews the educational records of all potential ELs identified through the questionnaire, identifies those with possible ESL needs, and refers them for assessment and evaluation. The ESL teachers and general-education teachers determine whether any academic problems a student is having are related to his or her English-language proficiency based on the assessment results and student performance in class. If so, the student is referred to ESL services.

Similarly, DoDEA schools have an elaborate and clear identification process for students with special needs. This process is in accordance with the Individuals with Disabilities Education Act (Pub. L. 101–476, 2010). Students with physical disabilities or known cognitive handicaps are usually identified upon enrollment. Other students with potential special needs tend to be identified by the teacher because they are having difficulty learning; teachers refer such students to the SST, which develops specific interventions for them. If students do not improve, they are then referred to a case-study team, which reviews students' portfolios and determines whether students need to be tested for a more severe cognitive disability. The special-education program targets students with mild, moderate, and severe disabilities covering a range of conditions, such as learning disabilities, communication and emotional impairment, and development delay. The availability of ESL and special-education staff varied across schools, with some schools indicating a need for more ESL and special-education teachers to provide the needed services.

DoDEA prioritizes the inclusion of students with special needs and ELs in general-education classes and the broader school community to the extent possible. For students with special needs, DoDEA policy also provides other options, such as pulling students from class and providing them with services in separate resource rooms for a portion of the day or, in severe cases, at home or in off-campus facilities. Most district ISSs raised concerns about how the inclusion policies were implemented in the classrooms. They indicated that teachers were not implementing differentiation techniques sufficiently to reach special-education and low-performing students. One ISS noted, "There is little differentiation going on. They teach to the middle of the group. The differentiation only comes in the form of retest or extra time." Some ISSs also indicated that it is difficult for teachers to find time to differentiate when they are expected to cover "too many standards." Our classroom visit data showed differentiation occurring in only 15 percent of the lessons observed. In these classrooms, teachers provided opportunities for students at different proficiency levels to engage in tasks or activities at their learning levels.

For example, in one class, students were working an activity called Math Message. The message asked, "If student A measured an object in centimeters, and student B measured it in inches, who do you think will report the larger number?" (Students were not given rulers or an object but were asked to think about the math message.) The teacher realized that four students did not understand the problem and pulled them aside to show them another way of thinking about the problem: by visualizing the difference between inches and centimeters when measuring a box. She said, "Before you do anything, let's do an experiment. Measure this box using the centimeter ruler: How long is it?" The student said, "32." Then she said, "Let's do it in inches." The student said, "It's 13 inches." Then the teacher said, "So, let's go back to the question: Who do you think will report the larger number?"

However, the majority of classroom teachers made very few adjustments in assignments or resources or activities. All teachers observed provided one-on-one assistance when students struggled or asked for help. But they tended to revisit or reteach the mathematical concepts using the same materials and similar instructional strategies. Although our classroom observations indicated that not many teachers implement differentiation strategies, teacher survey self-reports showed otherwise. About one-third of surveyed teachers indicated that they plan different assignments for individuals or groups of students based on their performance at least once during a lesson.

Gifted Students

The identification of gifted students did not appear to be as consistent across schools as the identification of EL students and students with special needs. Usually the process is initiated by classroom teachers referring students to be tested for gifted status. The Gifted Review Committee reviews teachers' recommendations and assessment results and determines whether the student should be eligible to receive gifted services. Few schools mentioned the provision of gifted services. These schools indicated that gifted services were provided within the context of students' regular class or as pull-out activities on a periodic basis.

Teacher Content Knowledge

There is consensus in the literature that effective teaching requires strong content knowledge of mathematics. Of the surveyed teachers, 64 percent reported "understanding the details of DoDEA mathematics standards well enough to use them in development of lessons and assessments, as well as explain the standards thoroughly to colleagues." Similarly, approximately 75 percent of the surveyed teachers indicated that they were "very prepared in terms of their knowledge and skills to teach students the mathematics standards." Many principals we interviewed indicated that their teachers were familiar with the content of the DoDEA standards. Nonetheless, some principals, teachers, and ISSs perceived that teachers lacked adequate content knowledge, especially teachers in the lower grade levels. They raised concerns about teachers not understanding the concepts behind the mathematical processes. When talking about the qualification of teachers, one ISS said, "Any standard involving computation is easy to implement. Basic operation is easy to spend time on. Teachers do not know how to teach in depth. Teachers lack mathematic knowledge to be able to teach effectively." A few teachers and principals indicated that elementary school teachers are not fluent in mathematics because they do not have strong mathematics backgrounds or because they do not like mathematics. Perceptions about teacher knowledge in mathematics were more favorable at the higher grade levels, possibly because teachers at middle and high-school levels tend to have degrees in mathematics.

Classroom Assessments

Best practices in classroom assessment include having a thorough program of formative assessment that is aligned with standards, places equal emphasis on content and practices, and is supported by technology. Such classroom assessment activities complement summative assessments that are standardized across classrooms (and are aligned with the same standards). The formative classroom assessments provide teachers with immediate, proximal, lesson-related information about small increments in student learning that can be used to guide day-to-day instruction. The summative assessments, in contrast, provide annual (or periodic) measures of accumulated learning that allow administrators to assess the effectiveness of the mathematics program overall and to compare performance across schools and districts.

Our observations of DoDEA mathematics lessons and DoDEA teachers' survey responses both indicate that mathematics teachers regularly assessed student knowledge in class, although the type, frequency, and focus of these assessments varied considerably across teachers. We saw evidence in the observations and artifacts of many classroom-assessment strategies, including daily quizzes, unit tests, student interviews, student self-rating scales, daily exit cards, remote key pads ("clickers") linked to display software, and performance tasks to ascertain how well students have mastered mathematical content and practices.

Common Classroom Assessment Practices

Fully half of the mathematics lessons we observed included some formative assessment of student knowledge. In addition to pencil-and-paper tests and quizzes, we saw examples of teachers using warm-up problems at the start of lessons, white marker boards and electronic response devices ("clickers") during lessons, exit tickets at the end of lessons, and other techniques for measuring students' understanding of lesson content. Most of the assessment artifacts teachers provided were drawn from textbooks and other curriculum materials, but some of the classroom assessments came from other sources. For example, a few teachers told us that they used online resources to find assessments. In addition, a few schools had adopted periodic supplemental assessment activities to encourage particular aspects of mathematics. For example, teachers in a few schools were asking students to complete "exemplars," which were grade level–specific problem-solving exercises developed elsewhere to foster mathematical problem-solving.

Survey responses suggest that assessment was a regular part of teachers' classroom routines. Teachers indicated that they emphasized a number of assessment activities. The vast majority of teachers reported putting moderate or major emphasis on "giving tests and quizzes to find out what students know" (79 percent) and on "using informal questions to assess student understanding" (94 percent). We also asked about actual practice, and the results indicated that most teachers regularly assessed student performance in class. More than two-thirds of teachers reported that they "reviewed assessment results to identify individual students who need supplemental instruction" one or more times a week, and more than 25 percent reported doing so one or more times per lesson.

The assessments we collected as artifacts of instruction were far more likely to measure mathematics content than mathematical practices. Typical of the former were tests that focused on specific knowledge or procedures, such as the test shown in Figure 3.1. On the other hand, we did see instances in which classroom assignments were used to assess mathematical practices (e.g., processes and proficiencies). Figure 3.2 is an instance in which students were asked not only to graph information but also to interpret it in the context of a story or event. This example embodies an assessment that tries to measure a student's ability to reason abstractly and quantitatively (a key mathematical practice in the CCR standards).

We saw very little use of technology for assessment during our observations. The one possible exception was IXL Math, which offered students rapid, gamelike tests of skills and tracked their performance. Some teachers used IXL to give a group of students an opportunity to reinforce skills; however, we saw no instances in which teachers were monitoring the results of the IXL measures. We were told that the contract to purchase access to IXL would not be renewed, and the use may have been influenced by the belief that the resources would no longer be available.

Variations in Classroom Assessment

In the classes we observed, DoDEA teachers regularly used classroom assessment to find out what students had learned. Yet, there was considerable variation in the assessment strategies they used. In the lessons we observed, we saw students assessed using tests, problems, or tasks drawn from a huge range of sources, including Exam View, Schoology, Center for Science Industry test (CSI) tests, IXL Math, Scholastic Mathematics Inventory, Exemplar, a commercial product referred to as the "diagnostic" test, Teachers Pay Teachers, Pearson online, STAR Math, Glencoe, school-constructed common pre and post tests, and others. It is not necessarily bad to find variation in classroom assessment practices. Yet, in the context of DoDEA, which has common mathematics standards and curriculum materials, the range of assessments seemed unjustified. The tests and measures we observed varied among grades within schools, among schools within districts, and among districts.

Figure 3.1
Assessment Focused on Mathematics Content

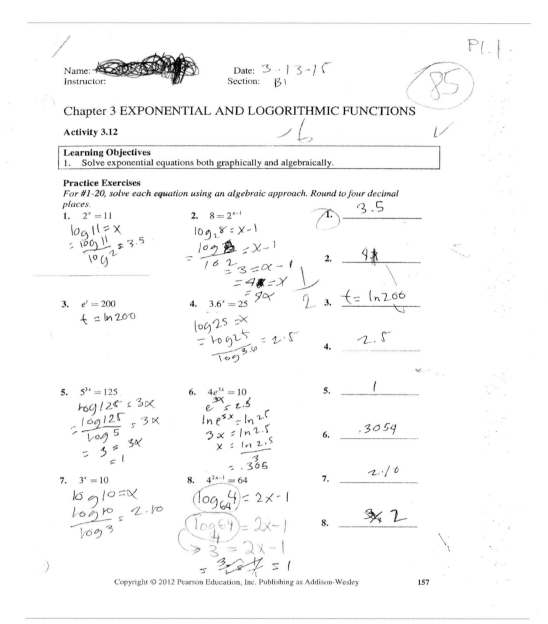

RAND *RR1272-3.1*

Professional Development for Assessment

On the survey, many teachers reported that classroom assessment was a focus of their PD during 2014–2015, and many reported needing more PD related to assessment. Just more than half of the teachers surveyed (58 percent) indicated that "use of assessment data to inform instruction" was a moderate or major focus of their PD during 2014–2015, and 31 percent reported that they had a moderate or high need for additional PD on this topic. Similarly, more than one-third (38 percent) reported that their PD had a moderate or major emphasis on "developing classroom assessments," while 43 percent reported a moderate or high need for additional PD on this topic.

Figure 3.2
Assessment Focused on Mathematical Practices

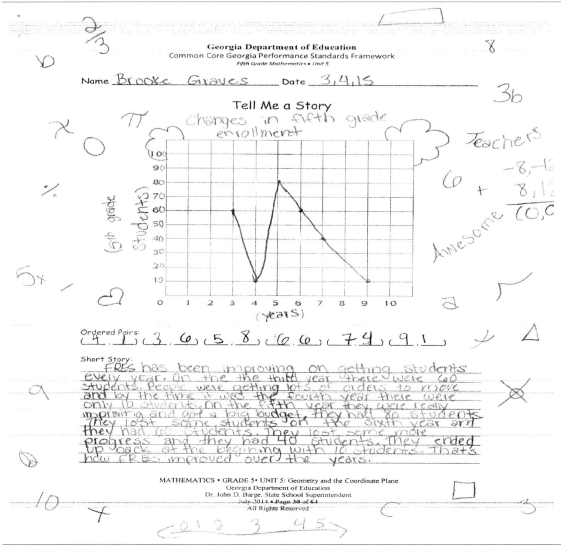

Large-Scale Standardized Testing

As one might expect, we observed a wide range of formative classroom assessments, but teachers mentioned only one or two annual, summative assessments. DoDEA mandates annual standardized testing in reading and mathematics for students in selected grade levels, using the TerraNova, PSAT/NMSQT, and SAT tests. Principals we interviewed used results from those annual tests to monitor their schools' overall performance over time and to inform CSI goal setting. Teachers we interviewed generally indicated that they made efforts to prepare students to take those tests because the results had important consequences for their schools. However, they did not think that the tests were well aligned with the current DoDEA standards and did not know whether they were aligned with incoming the CCR standards. As one teacher put it, "It's frustrating. DoDEA standards in [my] grade may say one set of skills, but that's not what's on the TerraNova test. So it's not valid." Most teachers reported that they did not use the results from the annual standardized tests for daily instructional improvement. We did see instances in which the results from the standardized tests were displayed on bulletin boards as an indication of the school's progress over time and as evidence of achieving identified goals related to mathematics performance.

Instructional Leadership

In the research literature, instructional leadership is described as a multidimensional characteristic, and, not surprisingly, there are a number of different ways to characterize the skills and activities of effective instructional leaders. For the purposes of this study, we adopted a framework that describes best practices in this area in terms of three key actions. First, instructional leaders set direction, which entails creating a shared vision within the school with high expectations for student performance and communicating this vision to teachers, parents, and other stakeholders. Second, effective instructional leaders engage in practices to develop the capacities of their staff. This includes setting expectations for practice in terms of instruction, assessment, and collaboration. It also involves assessing staff needs in these areas and developing appropriate PD programs. Finally, to the extent necessary, instructional leaders redesign their organizations to support their vision and their people. This task involves strengthening the culture of the school around these goals and building collaborative processes that sustain the changes.

In the case of DoDEA, most of the school and district leaders we interviewed told us that DoDEA headquarters (HQ) is responsible for setting the direction as far as the mathematics program is concerned. As one superintendent put it,

> Policy and procedures came from headquarters; that's where we get our marching orders. And then the area follows through with that. It goes from headquarters to area, area to district, and district to principal. Each of us has a responsibility to support the mission.

Not all administrators had such a "top-down" perspective; nevertheless, most of the school leaders were waiting for DoDEA direction before planning the implementation of the CCR standards. For example, almost without exception, the principals we interviewed had not yet shared information about the new mathematics standards with parents. Most of our school visits occurred before principals or teachers had received any official training on the CCR standards (although the schedule for training was announced while we were conducting this study). It is not surprising that there had been little formal discussion of the CCR standards in the schools we visited. Yet, some school leaders had taken it upon themselves to start learning about the CCR standards in anticipation of the new standards. One or two school leaders had begun to explore the CCR standards by developing a crosswalk between the current DoDEA standards and the CCR standards or by examining the alignment of their current curricula to the new standards. In addition, in some schools, individual teachers were seeking out information about the CCR standards on their own or were familiar with the CCSS from previous teaching assignments in U.S. public schools. It is not clear to us whether these few initiatives came initially from the school leaders or the teachers, but they were being were encouraged by the principals in these schools.

During interviews, teachers and principals described a variety of things that instructional leaders do to enhance the mathematics program. The most-common efforts included engaging in learning walks and performance-appraisal visits, providing one-on-one coaching or mentoring, planning or obtaining PD for staff, fostering collaboration among staff, reviewing data on student performance and sharing it with staff, and providing curriculum resources to support instruction. These are all important aspects of staff improvement. Yet, schools varied considerably when it came to who provides instructional leadership for mathematics. We asked both teachers and principals who served as the instructional leaders for mathematics in their schools, and the results varied widely. In some instances, staff members said the principal was the instructional leader; in some instances, the mathematics ISS was identified as the leader; and in still other instances, staff identified individual teachers as the "go-to" people when it came to questions about mathematics teaching. A few respondents indicated that the school's mathematics support specialist (where this position existed), the mathematics department chair (in some high schools), the assistant principal, or the CSI team served as the instructional leader in mathematics. The de facto instructional leader was not always the official instructional leader. In many schools we visited, individual staff identified different people as the mathematics instructional leader within the same school. And in some cases, staff told us that they consulted with different people, depending on the nature of the question; one person might be the strongest resource when it came to technology, another when it came to teaching specific mathematics topics.

This finding from the interviews suggests that, in many schools, there did not seem to be one recognized mathematics instructional leader; instead there was an informal system of diffused leadership within the schools. For example, in some schools, mathematics teachers informally identified one or two experienced mathematics teachers at their schools or at the district level as the person(s) to call with problems pertaining to use of technology in mathematics classes or the implementation of specific instructional practices. Principals at those schools tended to make decisions regarding the purchase of technology or other resources to facilitate the implementation of the mathematics program. Principals who had strong mathematics skills were more likely to be identified (or to self-identify) as instructional leaders than principals who were trained in other areas. Mathematics ISSs were more likely to be identified as instructional leaders in situations in which they were easily accessible to schools (e.g., geographically close) and had established relationships with teachers than in situations in which they were more distant (e.g., were not available on "a day's notice") or were not well known by staff.

In the interviews, we asked specifically about the role of the mathematics ISS because this was often the only person with designated responsibilities related to improving mathematics teaching. Two different functional models emerged. In one, the mathematics ISS was seen as a leader who initiated contact with schools or teachers to try to improve mathematics instruction. In the other, the mathematics ISS was seen as a supporting resource who was available when called upon to offer individual assistance. Mathematics ISSs recognized that their role was a difficult one because they had no direct authority to influence schools or teachers but instead had to encourage people to ask for their help. Some told us that they tried to "push" themselves out to schools, whereas others waited to be "pulled" in by staff. Individual personalities may also play a role; some mathematics ISSs were perceived by teachers to be helpful resources, while others were perceived as unhelpful or unresponsive.

Finally, although it is difficult to measure organizational features in a single visit, it seemed clear from our interviews that schools had very different professional learning cultures and that school leaders supported teachers in different ways. For example, some schools promoted active professional learning communities among teachers, and some did not press for this type of collaborative engagement. Some schools appeared to be highly collaborative when it came to mathematics teaching and learning (teachers met to discuss mathematics teaching, shared lessons, called upon one another if there were problems), and other schools were much more individualized (teachers were responsible for their own lessons and might reach out to an instructional leader on their own if they had questions). We were not able to tell the extent to which the school leader had acted to create a particular culture for the school, but we did observe that, in some instances, communication about mathematics could be described as a web of interconnections, while, in other cases, there was more of a hub-and-spoke arrangement.

Administrator and Teacher Professional Development

Research suggests that effective PD for implementing the CCR standards should do the following:

- Be sustained and intensive.
- Be content focused.
- Involve active learning on the part of teachers, integrated into daily activities.
- Involve collective learning.
- Focus on the major shifts inherent in the CCR standards.

Many of these features were mentioned during our interviews with administrators in the three U.S. LEAs that were leaders in implementing the CCSS. In particular, they stressed the importance of sustained, content-focused training (over multiple years), an emphasis on the new standards, and the use of active learning rather than lecture.

Based on our interviews with staff, it does not appear that recent DoDEA PD in mathematics incorporated many of the identified features of high-quality PD. Most of the teachers we interviewed reported receiving very little PD in mathematics during the past year, so the comments they did make reflected their experiences from prior years; in some cases, teachers referred to more-extensive PD they had received up to ten years ago. Although teachers continue to receive some training and PD, it appears that it is more likely to be of short duration, not connected, and not have the focus of these more-memorable prior PD experiences. When describing past PD experiences, the teachers we interviewed reported few instances of PD that were sustained, active, or focused on building a community of expertise.

It appeared to us that the local capacity for delivering high-quality mathematics PD or supporting mathematics professional learning communities was very uneven. Some school leaders described clear plans for PD to meet teachers' needs, while others were imprecise or uncertain. There appeared to be greater subject-matter expertise in middle and high schools than in elementary schools, offering stronger resources on which to build local PD.

We should also note that survey responses suggested that more than half of responding teachers received some PD in 2014–2015; 60 percent of teachers in DoDEA Europe and DoDEA Pacific reported receiving PD in 2014–2015, as did 82 percent of teachers in

DoDEA Americas. Survey respondents indicated that their PD focused on a variety of topics, ranging from content of the mathematics standards to differentiation of instruction for students with disabilities; each of the nine topics we asked about was a major focus of PD for 10 to 30 percent of the teachers. We cannot tell how long these PD activities lasted or how well they were received, but, as noted in the next paragraph, the majority of survey respondents indicated that they needed additional training on a number of the topics.

We also asked teachers in the online survey about their needs for PD across a wide range of topics. About three-quarters of the teachers said that they had a moderate or high need for PD focused on "understanding the difference between mathematics concepts addressed by the CCR standards for mathematics and those addressed by the DoDEA mathematics standards" and "understanding which mathematics concepts are being prioritized in the CCR standards for mathematics." These needs were far more widespread than needs for PD related to specific mathematics topics, such as making connections, using mathematical language, or developing conceptual understanding. This suggests that teachers are ready to receive the planned initial training on the CCR standards. After that training is complete, teachers may voice greater needs for additional PD around specific topics.

STEM Opportunities

The educational experts suggested that effective STEM learning opportunities have the following characteristics:

- Incorporate STEM professionals as role models.
- Involve partnerships with STEM-rich institutions.
- Ask students to engage in doing STEM-related activities, not just hear and talk about them.
- Provide regular, consistent opportunities.
- Include cross-grade opportunities to interact with other students.

In most of the schools we visited, principals described some STEM enrichment opportunities, although few, if any, of them included all the features previously described. After-school clubs that met on a regular schedule were the most common way in which schools offered students enrichment related to STEM. Principals described clubs related to robotics, computer coding (e.g., Cyber Patriots), biology, science, hydroponics, engineering (including Lego), and other topics. In addition to STEM clubs, many schools offered STEM events that occurred once or twice a year. These events included science fairs; engineering field trips; Earth Day; science, technology, engineering, arts, and mathematics night; gardening; a navigation program; "this is what I do" events with military STEM personnel; cup stacking; and Mathematics Olympiad. Some schools offer special STEM activities that are more closely integrated into the school day (but do not substitute for regular mathematics and science instruction). For example, one school has a STEM lab that students attend for an hour a day every fourth week, and other groups of students rotate into the lab during the alternate weeks. A few schools had no supplemental STEM activities.

Looking across this diverse set of activities, we see many instances in which the school uses the expertise of local military personnel and facilities (e.g., hospital personnel, U.S. Air

Force resources). In addition, according to principals and teachers, all the activities appear to involve active learning by students, i.e., doing STEM-related activities, not just listening to people talk about them. Most clubs met regularly, so the opportunities were lasting, although we did hear about cases in which the club lasted only a few sessions. There were fewer examples of partnerships with nonmilitary organizations.

Based on the interviews, it appears that students' access to STEM opportunities varied considerably depending on the schools where they were enrolled. Key factors in access to STEM enrichment seemed to be proximity to STEM-rich organizations, staff expertise, and local capacity to support supplemental STEM materials and programs. We heard about more STEM enrichment at the elementary level than at the secondary level, but more students at the secondary level were already enrolled in science or technology courses.

We heard varying comments from principals about the value of STEM-related enrichment activities. On the one hand, some principals embraced the idea and had developed strong connections with local STEM professionals. On the other hand, a few principals told us that they thought that STEM should be integrated into the curriculum, not separated from it. One reported that the whole exercise of creating STEM clubs was a compliance activity they did to check off a box for DoDEA HQ.

Summary

Implementing a high-quality mathematics program that incorporates best practices in each of the eight key programmatic areas is a challenging endeavor. Our results suggest that DoDEA has been successful in incorporating some best practices in its mathematics program, but there is room for improvement. Teachers reported a desire for more curriculum resources on which they could draw to support standards-based instruction, and that will be important with the implementation of the CCR standards. The same is true of both standardized and formative classroom assessments. Mathematics curriculum and instruction appeared to be widely focused on standards but not always as coherently, deeply, or rigorously as would be optimal. Schools attended to the needs of various students groups, but teachers desired help in improving their skills in differentiating instruction. Efforts were made to provide supplemental STEM activities, but they varied widely based on local resources. Schools did not always have strong instructional leadership for the mathematics program, nor did they provide the highest-quality PD to support teacher improvement in mathematics, although teachers believe that they have the content knowledge needed to be effective mathematics instructors. In the next chapter, we focus on the question of implementation of the CCR standards and share observations about how DoDEA might implement the new standards in the most-effective ways.

CHAPTER FOUR

Concerns About the Upcoming Implementation of the College and Career Readiness Standards

In the course of studying the current DoDEA mathematics program, we learned a number of things about DoDEA's efforts to implement the existing (2009) standards and support high-quality mathematics curriculum and instruction. Teacher and principal comments about the DoDEA mathematics standards and existing supports for mathematics teachers suggest challenges that will need to be overcome in order for the upcoming implementation of the mathematics CCR standards to be successful. This chapter identifies a few potential problems that DoDEA may want to address to ensure that the CCR mathematics standards are implemented well and that the mathematics program is as effective as possible in the future.

We note that our interviews and observations occurred after the adoption of the CCR standards had been announced but prior to principals or teachers receiving any formal training on them. We collected data during the time that the training program was being developed and the initial training schedule was announced. Thus, most teachers and administrators were aware that new mathematics standards were coming, many had already heard something about the CCSS and the controversies associated with it, some had taken the initiative to learn more about the standards on their own, and, in a few of our later interviews, administrators had already participated in introductory training sessions. As a result, we heard a range of things about current and past DoDEA efforts to improve the mathematics program. However, we did not systematically query all the teachers and principals we interviewed about their prior training and support, so we do not know how prevalent each of these attitudes or opinions is among DoDEA staff. To be conservative, we do not report concerns expressed by only one or two individuals. Instead, we focus on a few noteworthy problems that might plague the implementation of the CCR standards in mathematics based on reports from a number of respondents.

Training and Support for the Mathematics Program

A number of teachers raised concerns about the mathematics PD they had received previously from DoDEA. (At the time of our interviews, they had not yet participated in any CCR standards training.) Teachers complained that PD was too infrequent and that it was not equally available to all teachers. Teachers in more–geographically remote schools told us that the PD they received, which was delivered remotely through video (even two-way interactive video), was not as effective as PD delivered in person. DoDEA teachers also said that previous improvement initiatives were not sustained long enough but were superseded too quickly by new directions. DoDEA high school teachers and administrators made similar comments about training that never reached them: "By the time reforms are slated to get to the high schools, they've been changed and we don't get anything." How much training should DoDEA provide to support the implementation of the CCR standards? The exemplary LEAs we contacted indicated that their efforts to implement the CCSS in mathematics included three or more years of PD and support for teacher change. They also reported involving teachers in all the elements of implementation—from interpreting the standards to identifying appropriate curriculum materials to designing training activities. They reported that changes of this magnitude take time, and they suggested that it would not work to shortchange the process by trying to implement faster.

By the time we completed our interviews, DoDEA had provided the initial PD to administrators. Superintendents with whom we spoke near the end of the study had seen the plans for principal and teacher PD beginning in summer 2015, and they made positive comments about the amount of training being provided, the phased implementation strategy, and the plans for follow-up for teachers throughout the year. They praised DoDEA's systematic approach to implementing the CCR standards and noted that current efforts appeared to be better organized than past efforts.

The Role of the Mathematics Instructional Support Specialist

The mathematics ISSs play a key role in DoDEA's mathematics program, offering first-line support to teachers, who have questions about content, curriculum, and instruction. Teachers had mixed opinions about the effectiveness of the ISSs. Although some teachers praised the support they received from their local mathematics ISSs, many teachers were critical of interactions with their mathematics ISSs. They faulted ISS support for not being responsive (taking too long), not being informative (lacking useful mathematics content or instructional insights), or not being engaging (lacking rapport). There was considerable variation in attitudes toward mathematics ISSs, and some of it seemed to be a function of the personality and skills of individual ISSs. To support the implementation of the CCR standards, which emphasized deeper understanding of mathematics, the ISS will need strong mathematics content knowledge. While some principals we interviewed believe that it is essential for mathematics ISSs to have strong mathematics skills, others believe that a person with good staff development skills can fill this role if he or she has good training materials. A few teachers reported that, in their experience, their local school mathematics support specialist is likely to be a person with general training skills, not necessarily a person with strong understanding of mathematics and mathematics pedagogy.

There were also differences in expectations about how ISSs should function. Some ISSs told us that they wait to be called on rather than actively pursuing teachers because they worry about overstepping their role. In contrast, one or two ISSs said they saw it as their job to reach out to teachers and see whether they were delivering good mathematics instruction.

A final challenge we identified is how to support improvement of mathematics teaching in more-remote schools. The theme of remoteness came up in interviews with ISSs, principals, and teachers, particularly in geographically isolated places. In these locations, teachers were more likely to report that they were not visited regularly by an ISS and that they needed more-sustained, classroom-based support from instructional leaders. Similarly, some ISSs were frustrated that they could not provide that level of support. They noted that building a productive, trust-based relationship takes time and face-to-face contact, which is not possible in some locations.

Alignment with the College and Career Readiness Standards

It is important to align operable standards, curriculum materials, formative and summative assessments, and instructional support. Most teachers understood this; for example, they indicated that the textbooks should be aligned with the standards, and they recognized that the books being used were selected with the DoDEA standards in mind, not the CCR standards. They were concerned about what would happen after the implementation of the CCR standards: Would there be new textbooks, or would they have to figure out how to assemble lessons from various sources to teach the CCR standards? Teachers also were concerned that any change in standards (and hence in teaching) would jeopardize their students' performance on standardized tests—the SAT in the case of high schools and the TerraNova for lower grades. They worried that teaching to the new standards would lessen student performance on the current tests, which are used for school accountability purposes. Adopting a new assessment that is aligned with the CCR standards would address this concern, and we understand that process is under way.

In most of the schools we visited, teachers expressed a desire for a common scope-and-sequence document that would make standards-aligned instruction more consistent across schools. This consistency would better serve the needs of students who frequently change schools.

Maintaining Mathematics Teacher Quality

Many principals told us that DoDEA personnel requirements made it difficult to hire high-quality mathematics teachers. Administrators complained to us that hiring rules interfered with their efforts to hire the most-qualified mathematics teachers. Both the centralized application process and the absolute veterans' preferences were mentioned as obstacles to effective hiring, particularly in mathematics and STEM fields.

Summary

Although the main focus of the study was to judge the quality of the current DoDEA mathematics program along eight dimensions in preparation for the implementation of the CCR standards, our interviews and observations revealed a number of issues that we believe will be relevant to successful implementation of new mathematics standards. These issues included training and support for mathematics teachers; the role of the mathematics ISS; alignment of the standards, curriculum, assessments, and PD; and obstacles to hiring effective teachers. We incorporate these ideas into our recommendations in the next chapter.

CHAPTER FIVE

Summary, Recommendations, and Next Steps

The purpose of our study was to examine the quality of the current DoDEA mathematics program in eight key areas and consider what those findings mean for DoDEA's plans to implement the CCR standards. In this chapter, we summarize our key findings in each area, as well as our observations regarding emerging efforts to implement the CCR standards. We offer a few specific recommendations to address issues we identified, and we conclude by suggesting three questions whose answers might help DoDEA assess and support implementation during the next few years.

Our study was conducted at the same time that DoDEA was planning the implementation of the CCR standards in mathematics. As we write this report, training and dissemination efforts are well under way. For example, every teacher in grades pre-K–5 would have already received a full day of training in August 2015, and four sessions of supplemental training were offered during the 2015–2016 school year. As a result, some of our findings may have been overtaken by events that were being planned as we conducted the study. Nevertheless, we summarize our findings, present our recommendations, and offer some thoughts about next steps based on the interviews we conducted; classrooms we observed; artifacts we collected; and, to a lesser extent, responses to the online teacher survey during the 2014–2015 school year.

Summary

Our results suggest that DoDEA has been successful in incorporating best practices in its mathematics program in some areas more than others. For example, DoDEA schools have incorporated many of the best practices identified in the literature in the areas of equity and diversity and classroom assessments. However, schools varied in their implementation of best practices related to curriculum resources, curriculum and instructional quality, STEM opportunities, and instructional leadership. There had been little recent administrator and teacher PD at the time of the site visits, not consistent with best practice. Details are presented in the following paragraphs.

Curriculum Resources

Many teachers indicated that DoDEA textbooks were not well aligned with the 2009 DoDEA mathematics standards or did not provide sufficient lesson alternatives to meet the needs of their students. Teachers reported not having standards-aligned curriculum resources readily available. Hence, teachers sought additional resources on their own to incorporate into their lessons because there was no system in place to help teachers identify and assess the quality of the resources.

Curriculum and Instructional Quality

The extent to which teachers implemented teaching practices that promote mathematical proficiency varied. In most lessons we observed, teachers made connections between mathematical concepts in the classroom but did not explore them in depth or provide opportunities for students to discuss them. Connections between mathematics and other subjects were less frequent. Although most teachers engaged in activities to promote understanding, procedural skills, and applications, we observed teachers balancing the three areas in less than half of the classrooms observed. Teachers seemed to be most comfortable emphasizing mathematical procedures. Nevertheless, some lessons we observed incorporated challenging activities designed to promote deeper understanding of mathematical concepts. These activities addressed authentic settings for which students were asked to make assumptions and predict future outcomes. All schools were well equipped with tools, manipulatives, and technology, but the extent to which they were used to promote mathematical understanding differed. Finally, we observed few teachers acting as facilitators to students who were working to develop their own solutions.

Equity and Diversity

All schools visited had systems in place to examine student performance and identify low-performing students. On one hand, schools offered services to low-performing students, including additional instruction time or support from mathematics specialists. On the other hand, teachers lacked curriculum resources that provide opportunities for low-performing students to work with visual presentations of mathematical ideas or computer-adaptive mathematics software that regularly assesses student mathematics knowledge, tracks student growth, and helps guide instruction. Schools also had effective systems for identifying ELs and students with special needs, but the identification of gifted students was more informal. Almost all schools visited implemented inclusion policies, but such policies were not accompanied by differentiation in instruction in the classroom to ensure student access to the curriculum.

Content Knowledge

There was widespread perception that teachers were familiar with the content of the current DoDEA standards but that they were not familiar with the CCR standards. There were some concerns that teachers, especially at the elementary grade levels, lacked in-depth content knowledge of mathematics or confidence in their knowledge. Principals reported that they had difficulty hiring teachers with strong mathematics backgrounds when they had vacancies and blamed DoDEA hiring policies for some of the problem.

Classroom Assessments

Teachers used a wide range of assessment techniques to monitor student progress. Although the frequency and type of classroom assessment varied considerably across teachers, content-focused assessment and assessments measuring procedural fluency were more prevalent than assessments that tried to measure mathematical practices (e.g., processes and proficiencies). Teachers made efforts to prepare students for the required summative tests, and that influenced the ways they assessed students within their classrooms. Teachers indicated that they would need training on assessments that are aligned with the CCR standards when these standards are adopted.

Instructional Leadership

DoDEA HQ has a strong voice in setting the instructional direction for schools in mathematics, and a number of administrators reported that their role was to carry out directions from HQ. It appeared to us that principals and teachers shared some responsibility for instructional leadership in mathematics daily. Principals reported engaging in learning walks, obtaining PD for staff, and fostering collaboration among staff. However, principals did not always self-identify and were not always identified by teachers as the instructional leaders in mathematics. In many schools, mathematics instructional leadership fell informally to other individuals, including the mathematics ISS, department chair, or individual teachers, depending on staff capacities and expertise.

Administrator and Teacher Professional Development

Principals and teachers have received very little extended PD pertaining to mathematics since the adoption of the 2009 DoDEA standards. Most reported a need for future PD to prepare them for the CCR standards and, to a lesser extent, to understand specific mathematics topics.

STEM Opportunities

Most schools provided some STEM enrichment opportunities to their students, including STEM clubs, STEM events (e.g., science fairs) and STEM activities integrated into the school day (e.g., STEM lab). However, student access to STEM opportunities and coverage varied considerably depending on staff expertise, resources, and proximity to STEM-rich organizations.

Recommendations

Move Quickly to Align Mathematics Curriculum, Assessments, and Support Services with College and Career Readiness Standards

There are a number of steps that need to be taken to align all elements of the mathematics program with the CCR standards, and DoDEA is well aware of the major challenges involved in aligning textbooks, assessments, and continuing training and support. We do not know the status of existing assessment and textbook contracts, but we know that school systems usually take many months to study alternative curriculum materials or assessments before deciding to make changes, and purchasing procedures can add additional time to the process. One key intermediate step DoDEA might take is to provide more curriculum resources now, before adoption of new textbooks and assessments. These resources could include a scope-and-sequence document or a pacing guide to help teachers match lesson content from current resources with grade-level standards in the CCR standards. Another step might be to provide more online resources, including possibly online individual coaching and mentoring, or an online forum for teachers to support one another with lesson materials and instructional ideas. Below we highlight a few actions that might not be as obvious but are potentially valuable according to the evidence we collected during this study.

Provide Training to Teachers in Locating Online Resources

Many of the teachers we interviewed indicated that they were using online resources to supplement textbooks and other materials available from DoDEA. It would be ideal for DoDEA to ensure that its teachers are wise consumers of online instructional resources. As U.S. states and districts move to implement the CCSS, more resources are being made available online. In some cases, such as in New York, newly developed lesson materials are prepared in electronic versions freely accessible to others. Within DoDEA, mathematics teachers are preparing and sharing their own lesson videos designed to help students learn independently. All DoDEA teachers should be able to access such resources and should receive training on how to judge the quality and applicability of digital resources to meet their own instructional needs.

Help Districts Develop Messages for Parents

The exemplary LEAs we interviewed provided parent workshops as part of their CCSS implementation efforts. They reported that it was important to keep parents informed about curricular changes, and it was critical to offer them opportunities to familiarize themselves with the skills their children would be asked to master and the kinds of lessons their children would be receiving. Parent workshops can help educate and deflect potential criticism, and schools should make efforts to inform their parent communities about the upcoming changes. Rather than expect each school to create its own parent-education program, it would be more efficient (and potentially more accurate) for DoDEA to prepare materials for parents and to develop workshops to help schools communicate about the CCR standards in mathematics.

Support Teachers' Efforts to Change Practice by Reducing Test-Based Accountability Pressures

Many teachers reported to us that they felt pressure to prepare students to do well on the TerraNova, the PSAT/NMSQT, and the SAT because results from these tests were used as indicators of school progress. Moreover, it was common for us to see one or two goal statements prominently displayed when we entered schools—improvement in mathematics outcomes and reading outcomes—along with test scores reflecting change over time. Some teachers told us that one of their concerns about the CCR standards is that teaching to the new standards would not prepare their students adequately for existing standardized tests. We cannot quantify the extent to which teachers' perceptions of test-focused accountability will undermine efforts to implement the CCR standards, but we consider it to be a real possibility. Under these circumstances, DoDEA might want to consider ways to deflect these pressures, particularly if there are plans to change assessments in the future. We cannot recommend a specific action without knowing more about future plans for standardized testing, but a number of options are possible; for example, it might be helpful to share examples of questions from CCSS-aligned assessments, such as Smarter Balanced or Partnership for Assessment of Readiness for College and Careers (PARCC), to help teachers understand the kinds of expectations DoDEA has for student understanding in mathematics.

Provide Time and Resources for Sustained, High-Quality Professional Development for Mathematics Instructional Support Specialists, Principals, and Teachers

DoDEA understands the importance of preparing staff to implement the CCR standards, and it is already making efforts to provide training for key staff at all levels of the system. At least on the surface, the plans that have been described to us seem to address most of the concerns that we heard from administrators and staff. We want to emphasize the importance of sustaining this support over multiple years to help teachers continue to improve their understanding of the standards and make changes to their teaching. The exemplary districts we interviewed found it necessary to continue to support teachers for three years to change their instruction to align with the CCSS. We do not know DoDEA's long-term strategy for PD related to mathematics, but we encourage the leadership to sustain the support provided to teachers for a minimum of two years and suggest that three years may be appropriate. It is harder to quantify exactly how much training should be provided each year, but the exemplary LEAs had teacher-leaders in every school who worked with their peers individually and in groups. Moreover, the teacher- leaders met regularly with experts to foster their own development.

Prepare Teachers to Differentiate Instruction to Meet the Needs of Individual Learners

Another area in which additional PD may be warranted is differentiating instruction for students with disabilities and students who are struggling with mathematics. Our observations suggest that teachers made adjustments during lessons to help students who did not understand the material, but they had not planned the lesson with alternatives in mind. As a result, their adaptations tended to be routine (e.g., explaining the procedure using similar words, repeating the same type of problem with smaller numbers) rather than using an alternative approach to make the material accessible. It is not easy to differentiate instruction in this manner, and teachers everywhere struggle with the challenges of meeting the needs of each student. DoDEA should assess the extent to which teachers struggle to meet their students' needs in mathematics and provide targeted support where necessary.

Prepare Teachers to Incorporate Technology in Their Instruction

It would also be beneficial if PD addressed ways in which technology could be incorporated effectively during classroom instruction. DoDEA invested significant resources to ensure that classrooms are equipped with up-to-date technologies, yet teachers used them in a rudimentary fashion. Future PD could include training teachers on how to integrate technology in their teaching to promote student mathematics discourse, increase student collaboration, and increase opportunities for students to express their understanding and present their new knowledge.

Build School Capacity to Support Mathematics Reform

Although centralized DoDEA support for the CCR standards is vital, no reform will be successful unless it brings about sustainable changes at the school level. School-based support is important for a number of reasons. First, teaching is a highly "situated" activity, i.e., it depends a great deal on the skills and personality of a given teacher and the set of students and resources assembled in a specific classroom. Moreover, teachers learn from watching their peers teach and talking to them about instructional challenges. Thus, it is hard to improve teaching from a distance. Second, the improvement of educational practices is an ongoing task; it is not accomplished through a two-day training workshop or a couple of interactions with an expert coach or mentor.

School leaders usually fill the role of instructional leaders because they are close at hand. They are frontline experts who can bring their insights to bear directly on the teaching-learning interaction in a given classroom over time. However, many principals are not mathematics content experts, and there appear to be gaps in terms of instructional leadership for mathematics. Thus, we think that DoDEA should find ways to bring additional instructional leadership to teachers.

This might be accomplished in many ways. Some districts create mathematics teacher-leader positions in each school. For this to happen, there would have to be an assigned individual in each school who is knowledgeable about the CCR standards content and practice standards and adept at supporting teachers in lesson development. Although district ISSs may be able to provide such support, their geographic distribution is not optimal for providing the ongoing support needed for implementing new standards. DoDEA might want to consider supplementing the role of ISS with a mathematics specialist or mathematics teacher-leader at each school who will work closely with district ISSs in the provision of teacher support. Currently, a few DoDEA schools have such mathematics specialists on their staff. Where they do exist, principals and teachers were positive about the quality of support provided by the specialists and reported developing trusting relationships—the specialist was viewed as part of the community. Teachers frequently sought their input regarding instruction and invited them into their classrooms to observe and provide feedback.

In addition to assigning the CCR standards expert staff at the school, teachers should be provided with ongoing opportunities for PD. These could include mathematics-team meetings, where teachers address general issues regarding the CCR standards, as well as meetings that cover specific CCR standards topics. Another approach is to build professional learning communities from which teachers create their own support networks. Such opportunities allow colleagues to work together to improve their understanding and implementation of the CCR standards. In addition, having active and reliable support in the form of professional learning communities might help develop and maintain enthusiasm for the CCR standards at schools, even as leadership or other staff changes.

An alternative that is closer to DoDEA's current strategy is to have sufficient mathematics teaching expertise "on call" locally. DoDEA could consider modifying the use of mathematics ISSs to better meet local conditions; for example, allocating more mathematics ISS positions to geographically distributed districts or providing more travel resources so mathematics ISSs can spend more time in schools. These changes might be paired with efforts previously described to develop mathematics teacher-leaders or subject-matter specialists in each school. In addition, online resources, such as the Teaching Channel, can help teachers with specific lesson ideas.

We do not think that there is single best strategy that DoDEA should employ everywhere; instead the organization will probably have to evolve a multipronged approach that meets its unique circumstances. However it is accomplished, DoDEA should strive to find the resources to enhance support services available to individual teachers to implement the CCR standards in mathematics.

Investigate Possible Problems in DoDEA Hiring Practices and Reduce Obstacles to Hiring Teachers with Strong Mathematics Expertise

A number of principals in overseas locations reported that they had difficulty hiring the most-qualified staff, particularly staff with strong content knowledge in mathematics and other STEM fields. We suggest that DoDEA investigate further to uncover the extent of this problem and, if it is widespread, try to develop policies and procedures to address it.

Start Working with High Schools Now to Ensure Broad Support for Future the College and Career Readiness Standards Implementation

DoDEA has engaged in various strategies to inform schools about the CCR standards, but much of its communication efforts have targeted elementary schools. This is reasonable because the reform is designed to roll out in different years, starting with the lower grade levels. At the time of our visits, most high school teachers indicated that they had not heard much about the CCR standards other than that DoDEA had adopted it. It was clear from the interviews that high school teachers were concerned about how the CCR standards would affect their students' college preparation and overall mathematics programs. Some teachers at the high schools were skeptical regarding the CCR standards. They were not convinced that the reform would actually be extended to the high schools. One teacher recalled her experience with previous DoDEA reforms that were abandoned after several years of implementation, suggesting that this might also happen to the CCR standards. As a result, we suggest that DoDEA send clearer messages about its intentions, particularly at the high-school level, to promote buy-in and support prior to implementation. These messages should address the positive aspects of the CCR standards and how the standards will benefit high-school students, and they should try to counter concerns that the new standards might harm students' preparation for college. Teacher buy-in and commitment to the CCR standards are critical because teachers are the ones who must implement all curriculum and instructional changes.

Monitor the College and Career Readiness Standards Implementation and Outcomes

Previous research on school reform shows that level and quality of implementation are major determinants of outcomes (Datnow, Borman, and Stringfield, 2000; Fullan, 1991; Vernez et al., 2006). Hence, it is important to monitor changes in practices and strategies undertaken by schools and teachers in order to identify problems and act to correct them. It is not realistic to expect schools to implement the CCR standards flawlessly from the outset. Schools are likely to struggle and vary in their levels of implementation and success. We recommend that DoDEA consider establishing a system to regularly collect information on school practices, as well as academic and nonacademic student outcomes (e.g., student scores on the CCR standards–aligned assessments, student motivation, interest in mathematics). This could be done through surveys, interviews, online logs, or other methods that would provide information allowing DoDEA to monitor the CCR standards implementation, identify areas in need of improvement at the system level, and provide timely and targeted support. At the district and school levels, administrators and teachers could examine the data pertaining to their own schools to inform their practices and identify needs for site-specific support. Another advantage of establishing a monitoring system is that it would allow DoDEA to identify areas in which the reform has been successful. This information could be shared with districts and schools to promote buy-in as the reform expands to middle and high schools and broadens to include ELA.

Next Steps

In addition to these specific recommendations, we identified a number of questions that DoDEA might want to investigate as it continues to implement the CCR standards. Answers to these questions will help DoDEA determine its next steps to support full implementation of the new standards.

How effective is the 2015–2016 training and support in preparing teachers to implement the CCR standards in mathematics? As we understand it, the implementation strategy involves a full-day orientation, four quarterly refresher training sessions, online support forums, online school leader training, and ISS mentoring. It will be important to monitor these efforts and find out whether they are having positive effects on teachers' knowledge, attitudes, and behaviors. This could lead to immediate improvements to address any identified shortcomings or unanticipated needs.

What follow-on efforts are needed to support implementation in subsequent years? As we noted above, information gathered during the implementation phase is critical for making midcourse corrections to address problems, support growth, and set the stage for future reform. For example, finding and sharing success stories is a powerful strategy for encouraging reluctant teachers to begin to change their practices. The key to effective follow-through is good insights into the kinds of obstacles and challenges that confront teachers who are attempting to implement the CCR standards and the reasons for teachers' reluctance to make changes.

Do the adoption of the CCR standards and accompanying changes in curriculum and instruction lead to better student outcomes? Ultimately, the test of the reform will be whether it leads to better student outcomes, including higher achievement, graduation rates, and college and career preparation. It is important to start planning now so DoDEA will be able to answer these questions in the future. For example, it would be useful to administer an assessment that is aligned with the CCR standards at the start of the reform to serve as a baseline for future comparisons. We do not know how much flexibility DoDEA has to modify its current testing regimen, but it would not be necessary to test all students on the new standards for the purposes of monitoring; a random sample would provide a reasonable overall baseline. It is also important to monitor implementation in terms of classroom practices to be able to associate changes in outcomes with specific aspects of the implementation of the CCR standards.

What lessons can be learned for implementing the CCR standards in ELA? The experience of implementing the CCR standards in mathematics can inform planned efforts to implement the CCR standards in ELA in the coming year. It would be wise to monitor the mathematics implementation with this perspective in mind so those responsible for ELA can benefit from the current efforts. If different people are responsible for ELA implementation, which is likely the case, it would be unfortunate if they did not benefit from the lessons learned from mathematics.

Schools Visited

Table A.1
Schools Visited

Region	District	School Name
Americas	Georgia and Alabama	Fort Rucker Primary School
		Fort Rucker Elementary School
	Kentucky	Lincoln Elementary School
		Lucas Elementary School
		Fort Campbell High School
	New York, Virginia, and Puerto Rico	Antilles Elementary School
		Antilles Middle School
		Antilles High School
Europe	Isles	Lakenheath Elementary School
		Lakenheath Middle School
		Lakenheath High School
	Kaiserslautern	Ramstein Elementary School
		Ramstein Middle School
		Ramstein High School
	Mediterranean	Vicenza Elementary School
		Vicenza Middle School
		Vicenza High School
Pacific	Japan	Sasebo Elementary School
		E. J. King High School
	Korea	Osan American Elementary School
		Osan American Middle School
		Osan American High School
	Okinawa	Zukeran Elementary School
		Lester Middle School
		Kadena High School

Demographics of Schools Visited

Table B.1
Student Characteristics of DoDEA Schools Visited and Rest of DoDEA Schools: Elementary Schools, as Percentages

Characteristic	Visited	Not Visited
Gender		
Male	50.9	51.2
Female	49.1	48.8
Race and ethnicity		
White	48.1	47.5
Nonwhite and other	51.9	51.7
Special programs		
ELs	19.6	12.5
Special needs	13.8	13.3
Reading proficiency		
Grade 3	70.9	69.7
Grade 4	70.3	72.3
Grade 5	67.9	69.5
Mathematics proficiency		
Grade 3	71.1	71.6
Grade 4	61.4	65.9
Grade 5	68.6	69.8

SOURCE: 2013–2014 DoDEA school report cards.
NOTE: Percentages may not add to 100 due to rounding.

Table B.2
Student Characteristics of DoDEA Schools Visited and Rest of DoDEA Schools: Middle Schools, as Percentages

Characteristic	Visited	Not Visited
Gender		
Male	51.4	50.9
Female	48.6	49.1
Race and ethnicity		
White	40.5	46.4
Nonwhite and other	59.5	53.6
Special programs		
ELs	24.9	7.7
Special needs	9.3	11.9
Reading proficiency		
Grade 6	79.1	77.9
Grade 7	78.5	78.4
Grade 8	81.2	78.8
Mathematics proficiency		
Grade 6	69.0	68.5
Grade 7	78.1	75.1
Grade 8	78.7	77.8

SOURCE: 2013–2014 DoDEA school report cards.

NOTE: Percentages may not add to 100 due to rounding.

Table B.3
Student Characteristics of DoDEA Schools Visited and Rest of DoDEA Schools: High Schools, as Percentages

Characteristic	Visited	Not Visited
Gender		
Male	49.8	51.2
Female	50.2	48.8
Race and ethnicity		
White	45.7	45.8
Nonwhite and other	54.3	54.2
Special programs		
ELs	9.3	14.9
Special needs	9.7	8.9
Reading proficiency (grade 9)	83.5	82.7
Mathematics proficiency (grade 9)	76.9	78.0

SOURCE: 2013–2014 DoDEA school report cards.

NOTE: Percentages may not add to 100 due to rounding.

Artifact and Observation Rubric

Table C.1
Mathematics Lesson Artifact and Observation Rubrics

Context

Grade Level/Subject: _____ Number of Students: _____

Organization (whole group, centers, individualized): _____

Recent Math Content: _____

I. Clarity of Lesson Objectives

The teacher communicated clear mathematical learning goal(s) for the lesson.

If the teacher articulates mathematical learning goal(s) for the lesson, please write them here:

_____
_____

Low	Mid	High
The teacher communicates no mathematical learning goal for the lesson.	The teacher communicates mathematical learning goal(s) for the lesson, but a majority of the work does not focus on mathematical concepts or processes that are connected to these goals.	The teacher communicates mathematical learning goal(s) for the lesson. The majority of the work and activities support students' work toward these goals.
Artifact examples: *Lesson objective is not written into lesson plans.*	*Artifact examples:* *Lesson objective is written into lesson plans. However, most planned classroom activities do not help students work toward the learning goals.*	*Artifact examples:* *Lesson objective is written into lesson plans. Most planned classroom activities help students work toward the learning goals.*
Observation examples: *Lesson objective is not written on the board, verbalized by the teacher, or shared with students in some way during the lesson.*	*Observation examples:* *Lesson objective is written on the board, verbalized by the teacher, or shared with students in some way. However, most classroom activities do not help students toward the learning goals.*	*Observation examples:* *Lesson objective is written on the board, verbalized by the teacher, or shared with students in some other way. Most classroom activities help students work toward learning goals.*

Observation Rating (Circle one): Evidence:	Low	Mid	High
Artifact Rating (Circle one): Evidence:	Low	Mid	High
Overall Rating (Circle one):	Low	Mid	High

Table C.1—Continued

II. Lesson Structure and Coherence
The lesson is coherently and logically organized to support student conceptual understanding of mathematics.

Low	Mid	High
Lesson is not coherent or logically organized. Classroom activities are not clearly related to one another. Artifact examples: *Lesson plan does not contain a clear sequence of planned activities that help to develop a mathematical idea or set of related ideas.* Observation examples: *Board work is disorganized and incoherent, and it is difficult to follow the logical development of a mathematical idea or set of related ideas during the course of the class period.*	Lesson is coherent and logically organized. However, the lesson does not promote connections between mathematical procedures and larger mathematics concepts or ideas. Artifact examples: *Lesson plan contains a logical sequence of planned activities. However, these activities focus only on procedures and make no attempt to connect to larger mathematical ideas or concepts. For example, division of fractions is presented using a "flip and multiply" procedure. Or linear relationships are taught using "rise over run" mnemonic, but the conceptual ideas behind these procedures are not discussed.* Observation examples: *Board work or classroom activities are clearly organized. However, the focus of this classroom work is primarily on the development of skills, and no effort is made to connect skills to larger mathematics concepts or ideas.*	Lesson is *conceptually* coherent throughout; activities are related mathematically and build on one another in a logical manner. Lessons are organized to allow students to make meaningful connections between concepts and procedures. Artifact examples: *Lesson plans show evidence that activities are intentionally sequenced, and the sequence logically develops a mathematical idea or set of related ideas. For example, slope is presented and developed as a rate of change. The graph of a line is presented as a representation of all solutions to a linear equality.* Observation examples: *Board work or classroom activities are clearly organized and help to foster student understanding. Classroom activities are organized to develop connections between skills, procedures, and concepts.*

Observation Rating (Circle one): Evidence:	Low	Mid	High
Artifact Rating (Circle one): Evidence:	Low	Mid	High
Overall Rating (Circle one):	Low	Mid	High

Table C.1—Continued

III. Student Explanations

Teacher uncovers student thinking, explanations, or justifications (including students' misconceptions) about mathematical content and concepts.

Low	Mid	High
No evidence of teacher intentions or work to uncover student thinking, explanation, or justifications about mathematical content/concepts. Artifact examples: *Plans, handouts, assignments, or assessments may ask students to "show their work," but prompts are intended to elicit bounded, procedural responses.* Observation examples: *Any questions or activities are intended to elicit short, bounded, procedural answers verbally or on the board (e.g., "25," "divide by 8," "The formula for finding the area of a triangle is . . . ," or showing procedural work for solving a problem on the board). When students share work on the blackboard, they share only solutions and solution steps, and their ideas are not developed and refined through class discussion.*	Evidence of teacher work to uncover student thinking, explanation, or justifications about mathematical content or concepts. However, "the student proposes, the teacher disposes"; the teacher does not facilitate discussion to explore or build on students' ideas, when potentially valuable. Likewise, the teacher may correct students' misconceptions but does not promote or foster exploration of underlying causes of these misunderstandings. Artifact examples: *Plans, handouts, assignments or assessments contain prompts for students to explain their thinking. However, artifacts do not provide evidence that the teacher intends for students to respond to one another's ideas or build on them.* Observation examples: *At least some questions probe students to go beyond short, procedural answers and are intended to elicit explanations, reasoning, conjectures, and justifications. However, in the observations, teachers do not ask follow-up questions or ask other students to share their perspectives in order to clarify, elaborate, or build on students' initial responses. In the same vein, if students make errors, the teacher may correct those errors or prompt them to give a different answer. But the teacher does not ask questions prompting students to reconsider their misconceptions (e.g., "Do you agree or disagree?" "Will that always be true?" "Can you think of a counterexample?").*	Evidence of teacher work to uncover student thinking, explanation, or justifications about mathematical content/concepts. In addition, the teacher intends for students to explain and defend their ideas and reasoning or to respond to, clarify, or build on one another's ideas. This may include classroom work exploring the underlying causes of student misconceptions. Artifact examples: *Plans, handouts, or assignments consistently contain prompts that ask students to explain their thinking. In addition, artifacts provide evidence of the teachers' interest in uncovering underlying misconceptions or intention for students to defend their ideas or respond to one another's ideas.* Observation examples: *Teacher questions probe students to go beyond short, procedural answers to provide explanations, reasoning, conjectures, and justifications. In addition, teachers ask follow-up questions or ask other students to provide their perspectives in order to clarify, elaborate or build on students' initial response or uncover and explore misconceptions.*

Observation Rating (Circle one): Evidence:		Low	Mid	High
Artifact Rating (Circle one): Evidence:		Low	Mid	High
Overall Rating (Circle one):		Low	Mid	High

Table C.1—Continued

IV. Connections Between Concepts Within a Grade Level or Across Grade Levels

The teacher helps students connect current learning with other important mathematics within or across grade levels.

Low	Mid	High
No connections are made to other mathematics concepts (for example, to prior skills or knowledge); or connections are made that were inappropriate or incorrect. Artifact examples: *Lesson plans, written assignments, handouts, or assessments contain no tasks, prompts, or activities that encourage students to see connections to other mathematical ideas or concepts.* Observation examples: *During class discussion, teacher or students make no connections to other mathematics or make connections to other mathematics that are incorrect or unjustified mathematically.*	Teacher explicitly connects lesson content to other important mathematics, but there is no deep exploration of those connections. Artifact examples: *Lesson plans, written assignments, handouts, or assessments contain connections to other mathematics, but these connections are cursory.* Observation examples: *During class discussion, the teacher brings up connections (e.g., "Yesterday we did adding fractions with like denominators; today we will do subtracting fractions with like denominators"). However, the students do not initiate, develop, or explore those connections themselves.*	Teacher explicitly includes one or more connections between lesson content and other important mathematics, *and* the lesson activities purposefully engaged students in exploration of those connections. Artifact examples: *Lesson plans, written assignments, handouts, or assessments contain tasks and prompts that require students to make connections between concepts and to explain or articulate these connections.* Observation examples: *During class discussion, students initiate, develop, or explore meaningful connections to other mathematics.*

Observation Rating (Circle one): Evidence:	Low	Mid	High
Artifact Rating (Circle one): Evidence:	Low	Mid	High
Overall Rating (Circle one):	Low	Mid	High

Table C.1—Continued

V. Connections Between Different Disciplines
The lesson encourages students to connect mathematics to other disciplines.

Low	Mid	High
No connections are made to other disciplines; or connections were made that are inappropriate or incorrect. Artifact examples: *Lesson plans, written assignments, handouts, or assessments contain no tasks, prompts, or activities that encourage students to see connections to other disciplines.* Observation examples: *During class discussion, teacher or students make no connections to other disciplines or make connections to other disciplines that are incorrect or unjustified mathematically.*	Teacher explicitly connects lesson content to other disciplines, but there is no deep exploration of those connections. Artifact examples: *Lesson plans, written assignments, handouts, or assessments contain connections to other disciplines, but these connections are cursory.* Observation examples: *During class discussion, the teacher brings up connections. However, the students do not initiate, develop, or explore those connections themselves.*	Teacher explicitly includes one or more connections between lesson content and other disciplines, *and* the lesson activities purposefully engage students in exploration of that connection. Artifact examples: *Lesson plans, written assignments, handouts, or assessments contain tasks and prompts that require students to make connections to other disciplines and to explain or articulate these connections.* Observation examples: *During class discussion, students initiate, develop, or explore meaningful connections to other disciplines.*

	Low	Mid	High
Observation Rating (Circle one): Evidence:	Low	Mid	High
Artifact Rating (Circle one): Evidence:	Low	Mid	High
Overall Rating (Circle one):	Low	Mid	High

Table C.1—Continued

VI. Cognitive Challenge
The teacher engages students in cognitively challenging tasks and encourages students' productive struggle with those tasks (e.g., making sense of and solving unfamiliar problems, reasoning mathematically, describing mathematical patterns, or constructing mathematical arguments).

Low	Mid	High
No evidence of teacher intentions to engage students in cognitively challenging tasks. Artifact examples: *Plans or assignments do not contain any cognitively challenging tasks. Instead, plans and assignments focus on work in which students apply memorized procedures or work on routine problems.* Observation examples: *The teacher does not provide students with cognitively challenging tasks. For example, the teacher tells students exactly what procedure to use to solve a problem, sets up problems for students from the outset, or asks students to do problems that they may already know how to solve.*	Evidence of teacher intentions to engage students in one or more cognitively challenging task. However, no evidence or only momentary evidence that teachers encourage and support students to engage in productive struggle with these tasks. Artifact examples: *Lesson plans, handouts, assessments, or assignments contain cognitively challenging tasks, although no compelling evidence that the teacher plans for students to persevere in completing those tasks. E.g., homework could contain optional "challenge" problems, or lesson plans may propose cognitively challenging tasks without providing evidence that the teacher is thinking about appropriate scaffolding for the tasks without solving the problem for students.* Observation examples: *For a small portion of the lesson, the teacher provides students with some challenge in determining how to set up or solve a problem. The teacher pushes for quick solution of the problems, cutting short students' work to solve it or not giving them enough time to solve it.*	Evidence of teacher intention to engage students in one or more cognitively challenging tasks, as well as more than momentary evidence that the teacher encourages and supports students to engage in productive struggle with these tasks. Artifact examples: *Beyond plans and assignments that contain cognitively challenging tasks, some evidence that students are expected to persevere in completing those tasks (e.g., a cognitively challenging homework assignment required of all students, an assessment in class that includes cognitively challenging problems).* Observation examples: *The teacher provides students with some challenge in determining how to set up or solve a problem. Additionally, the majority of students persevere in trying to solve the problem beyond a brief moment, and some solve the problem, although the teacher may eventually intervene to solve the problem.*

Observation Rating (Circle one): Evidence:	Low	Mid	High
Artifact Rating (Circle one): Evidence:	Low	Mid	High
Overall Rating (Circle one):	Low	Mid	High

Table C.1—Continued

VII. Modeling with Mathematics

The teacher helps students apply mathematics to real-world contexts and to solve problems arising in everyday life, society, and the workplace.

Low	Mid	High
Students are not required to apply the mathematics they know to solve problems arising in everyday life, society, and the workplace. Artifact examples: *Lesson plans, written assignments, handouts, or assessments contain no problems or activities that require students to use mathematics to model situations that arise in everyday life.* Observation examples: *During class discussion, the teacher uses a story problem to illustrate a situation but does not involve students in actively working on the problem.*	Students are provided with some opportunities to apply the mathematics they know to solve problems arising in everyday life, society, and the workplace, but these opportunities may be superficial or inauthentic. Artifact examples: *Homework could contain "application" problems that are unrealistic or require little thinking on the part of the student (e.g., simple application of memorized procedures for a word problem that is little more than a procedural problem with words around it).* Observation examples: *At least some instructional time during the lesson is spent work on applied (real-world) problems, although that work is brief (5 minutes or less). Or application problems are unrealistic or require little thinking on the part of the student (e.g., simple application of memorized procedures for a word problem that is little more than an procedural problem with words around it).*	Students are required to apply the mathematics they know to solve authentic, realistic problems arising in everyday life, society, and the workplace. Artifact examples: *Homework contains "application" problems that are realistic and require students to thoughtfully apply mathematics to real-world situations or contexts.* Observation Examples: *Modeling problems comprise a substantial portion of the class (more than 5 minutes). Modeling problems are realistic and require students to thoughtfully apply mathematics to real-world situations. Modeling may include using a short word problem from a mathematics text; figuring out among four recipes the proportion of orange juice and water that makes a mixture more orangey; or figuring out which is the best phone-call plan among three plans representing a linear, proportional, and stepwise function.*

Observation Rating (Circle one): Evidence:	Low	Mid	High
Artifact Rating (Circle one): Evidence:	Low	Mid	High
Overall Rating (Circle one):	Low	Mid	High

Table C.1—Continued

VIII. Responsiveness to Diverse Student Needs

The teacher is responsive to diverse students' needs and proficiencies, including necessary accommodations for special needs students or ELs.

Low	Mid	High
The teacher does not provide differentiated opportunities for students to engage in tasks or activities. Artifact examples: *Lesson plans, handouts, or assignments give no indication of differentiated opportunities for students at different levels.* Observation examples: *The teacher gives all students in the class the same tasks or work, using the same approaches, without providing adaptations or scaffolding to help students at varying levels persevere in solving problems.*	The teacher provides at least one differentiated opportunity for students, although the majority of tasks are not differentiated or intended to address the needs of students at varying levels of proficiency. When tasks are not differentiated, the teacher does not respond to student needs or cues for additional support during the lesson. Artifact examples: *Lesson plans, handouts, or assignments provide plans for at least one differentiated opportunity for students at different proficiency levels (e.g., plans for additional challenging activities for some students or adaptations or additional support for those who are having trouble). However, the majority of lesson tasks are not differentiated.* Observation examples: *The teacher provides at least one differentiated opportunity for students, including—but not limited to—additional challenging activities for some students or adaptations or additional support for those who are having trouble. However, when students appear to be having trouble, the teacher is unwilling or unable to support those students.*	The teacher provides consistent differentiated opportunities for students at different proficiency levels to engage in tasks or activities. However, if some tasks are not differentiated, the teacher responds to student needs or cues for additional support during the lesson. Artifact examples: *The majority of tasks or activities in lesson plans, handouts, or assignments provide differentiated opportunities for students at different proficiency levels (e.g., plans for additional challenging activities for some students or adaptations or additional support for those who are having trouble).* Observation examples: *The majority of lesson activities or tasks provide differentiated opportunities for students, including—but not limited to—additional challenging activities for some students or adaptations or additional support for those who are having trouble. For example, a teacher may have students work at centers or may use technology that allows for individualized instruction. Even if the majority of tasks are not differentiated, the teacher is responsive to student cues for support if they are having difficulty with an activity.*

Observation Rating (Circle one): Evidence:	Low	Mid	High
Artifact Rating (Circle one): Evidence:	Low	Mid	High
Overall Rating (Circle one):	Low	Mid	High

Table C.1—Continued

IX. Appropriate Use of Tools
The teacher provides students with appropriate technology or other tools that support students' thinking, reasoning, and learning.

Low	Mid	High
The teacher did not provide students with technological or other tools intended to support student learning. Observation examples: *Calculators, computers, or manipulatives are not available.*	The teacher may have provided students with technological or other tools intended to support student learning. However, the tools do not clearly enhance students' thinking and reasoning and may serve as a distraction that compromises that thinking and reasoning. Observation examples: *Calculators, computers, or manipulatives are available, but students do not use these tools productively to solve problems.*	The teacher provides students with technological or other tools during the lesson intended to support student learning. The tools provided by the teacher are appropriate and accessible and clearly enhanced students' thinking and reasoning. Observation examples: *Calculators, computers, or manipulatives are available, and students used these tools productively to solve problems.*

Observation Rating (Circle one): Evidence:		Low	Mid	High
Artifact Rating (Circle one): Evidence:		Low	Mid	High
Overall Rating (Circle one):		Low	Mid	High

Table C.1—Continued

X. Student Engagement with Mathematical Content
The teacher makes an effort to support students' engagement with the mathematical content of the lesson, or there is evidence that classroom norms and participation structures promoting engagement have been previously established.

Low	Mid	High
There is differential engagement or participation in mathematical content, and no apparent efforts or established classroom norms to address this issue. Observation examples: *Teacher calls only on volunteers or does not check in with student groups or individual students. Some students are disengaged or marginalized, and this differential access is not addressed. Nor is there evidence of established classroom norms that support students to share their responses and ideas (e.g., expectations of students to share and contend with one another's responses).*	There is differential engagement or participation, but the teacher makes effort to engage students in the mathematical content of the lesson, although it is not clear whether there are established participation structures or classroom norms that support students to share their ideas. Observation examples: *Teacher attempts to involve all students on occasion by calling on nonvolunteers, and checking in with student groups, checking in with quiet students. But teacher attempts may appear sporadic rather than systematic. There is not clear evidence of established classroom norms that support students to share their responses and ideas.*	The teacher actively supports and, to some degree, achieves broad and meaningful mathematical engagement *or* what appear to be established participation structures and established classroom norms result in such engagement. Observation examples: *Teacher attempts to involve all students consistently by calling on nonvolunteers, and checking in with most student groups, checking in with quiet students. There is some evidence of classroom norms for students to share their responses and ideas.*

	Low	Mid	High
Observation Rating (Circle one): Evidence:	Low	Mid	High
Artifact Rating (Circle one): Evidence:	Low	Mid	High
Overall Rating (Circle one):	Low	Mid	High

Table C.1—Continued

XI. Formative Assessment
Questions, tasks, or assessments do not yield data that would allow the teacher to assess students' progress toward the learning goals.

Low	Mid	High
The teacher does not assess student progress toward learning goals through questions, tasks, or assessments. Artifact examples: *No evidence that the teacher intended to assess student understanding or progress during the lesson via informal questioning of all students, a short quiz or assessment, or another means. Homework assignments—on their own—do not constitute formative assessment unless you have clear indication that the teacher used homework to understand the extent to which students understood or were progressing toward learning goals for the lesson.* Observation examples: *No evidence that the teacher assesses student understanding or progress during the lesson via informal questioning of all students, a short quiz or assessment, or another means. A problem that launches an activity or a lesson (such as a "Do Now"–type problem—does not constitute formative assessment unless you have a clear indication that the teacher intends to use this problem to surface or summarize student misunderstandings and misconceptions.*	The teacher assesses or partially assesses student progress toward learning goals through questions, tasks, or assessments. Artifact examples: *Some evidence that the teacher intends to assess student understanding or progress during the lesson via informal questioning of all students, a short quiz or assessment, or another means.* Observation examples: *Evidence that the teacher assesses student understanding or progress during the lesson via informal questioning of all students, a short quiz or assessment, or another means. Students may use software that provides teachers with diagnostic information about student performance or understanding, but there is limited evidence of how teachers use this information.*	The teacher assesses students' progress toward learning goals through questions, tasks, or assessment. In addition to assessing student progress, questions, tasks, or assessments also pinpoint where understanding breaks down. Artifact examples: *Some evidence that the teacher intends to assess student understanding or progress during the lesson via informal questioning of all students, a short quiz or assessment, or another means. In addition, evidence that assessment is intended to help the teacher note common student errors or misunderstandings that might be further addressed through additional instruction.* Observation examples: *Evidence that the teacher assesses student understanding or progress during the lesson via informal questioning of all students, a short quiz or assessment, or another means. In addition, assessment of student understanding helps the teacher note common student errors or misunderstandings that could be further addressed through additional instruction. Teacher may use instructional tools, such as clickers, to monitor student understanding, or there may be a clear indication that teacher uses data generated by instructional software (such as IXL) to assess student understanding or progress.*

Observation Rating (Circle one):   Low   Mid   High
Evidence:

Artifact Rating (Circle one):   Low   Mid   High
Evidence:

Overall Rating (Circle one):   Low   Mid   High

Observation Ratings

Table D.1
Observation Ratings, as Percentages

Observation	Low	Mid	High
Clarity of lesson objectives ($N = 46$)	63.0	6.5	30.4
Lesson structure and coherence ($N = 47$)	6.4	53.2	40.4
Student explanations ($N = 47$)	53.2	29.8	17.0
Connections between concepts within a grade level or across grade levels ($N = 47$)	51.1	46.8	2.1
Connections between different disciplines ($N = 47$)	83.0	17.0	0.0
Cognitive challenge ($N = 47$)	44.7	25.5	29.8
Modeling with mathematics ($N = 47$)	51.1	21.3	27.7
Responsiveness to diverse student needs ($N = 47$)	51.1	34.0	14.9
Appropriate use of tools ($N = 47$)	17.0	44.7	38.3
Student engagement with mathematical content ($N = 47$)	25.5	38.3	36.2
Formative assessment ($N = 47$)	29.8	53.2	17.0

NOTE: Percentages may not add to 100 due to rounding.

Artifact Ratings

Table E.1
Artifact Ratings, as Percentages

Artifact	Low	Mid	High
Clarity of lesson objectives (N = 39)	48.7	20.5	30.8
Lesson structure and coherence (N = 38)	39.5	34.2	26.3
Student explanations (N = 39)	59.0	28.2	12.8
Connections between concepts within a grade level or across grade levels (N = 39)	76.9	20.5	2.6
Connections between different disciplines (N = 40)	90.0	10.0	0.0
Cognitive challenge (N = 38)	50.0	39.5	10.5
Modeling with mathematics (N = 37)	48.7	32.4	18.9
Responsiveness to diverse student needs (N = 39)	79.5	18.0	2.6
Appropriate use of tools	N/A	N/A	N/A
Student engagement with mathematical content	N/A	N/A	N/A
Formative assessment (N = 38)	63.2	26.3	10.5

NOTE: Percentages may not add to 100 due to rounding.

Survey Findings

1. Which mathematics standards are your students expected to meet this school year (2014–2015)? (_N_ = 699)

	Overall	Americas	Pacific	Europe
My students don't have to meet mathematics standards this year	0.6	0.4	0.0	0.9
DoDEA mathematics standards	95.4	97.4	94.8	94.4
CCR standards for mathematics (i.e., Common Core State Standards for mathematics)	2.0	0.9	3.5	2.3
I don't know which mathematics standards apply this year	2.0	1.3	1.7	2.5

NOTES: All data in this appendix are given as percentages. Percentages may not add up to 100 due to rounding.

2. How familiar are you with the mathematics standards that your students are expected to meet this year (2014–2015)? (_N_ = 678)

	Overall	Americas	Pacific	Europe
I am unfamiliar or only slightly familiar with the mathematics standards.	1.3	1.3	0.0	1.8
I have a general understanding of the mathematics standards, but I am not familiar enough with the details to use them in the development of lessons and assessments.	5.5	4.0	6.3	6.2
I understand the details of the mathematics standards well enough to use them in the development of lessons and assessments but not well enough to explain the standards thoroughly to colleagues.	28.9	26.7	24.1	32.0
I understand the details of the mathematics standards well enough to use them in the development of lessons and assessments, as well as explain the standards thoroughly to colleagues.	64.3	68.0	69.6	60.1

NOTES: All data in this appendix are given as percentages. Percentages may not add up to 100 due to rounding.

3. How prepared are you—in terms of your knowledge and skills—to teach your students the mathematics standards they are expected to meet? (N = 677)

	Overall	Americas	Pacific	Europe
Not at all prepared	0.6	0.9	0.0	0.6
Slightly prepared	3.1	2.7	1.8	3.8
Moderately prepared	21.7	22.6	20.7	21.5
Very prepared	74.6	73.9	77.5	74.1

NOTES: All data in this appendix are given as percentages. Percentages may not add up to 100 due to rounding.

4. In 2014–2015, how much do you emphasize the following instructional strategies to help students become mathematical thinkers and master the mathematics standards? (N = 600–602)

		None	Minor	Moderate	Major
Asking students to solve unfamiliar problems that require mathematical thinking	Overall	0.8	9.0	43.2	47.0
	Americas	0.0	8.8	39.4	51.8
	Pacific	0.0	5.9	52.5	41.6
	Europe	1.6	10.1	42.5	45.8
Having students practice computation to develop fluency	Overall	2.0	13.8	35.4	48.8
	Americas	1.6	9.8	33.2	55.4
	Pacific	2.0	17.8	33.7	46.5
	Europe	2.3	14.9	37.3	45.5
Making connections between key mathematical concepts at my grade level(s) or in my course(s)	Overall	0.7	7.6	37.4	54.3
	Americas	0.0	9.3	32.1	58.6
	Pacific	0.0	5.9	41.6	52.5
	Europe	1.3	7.1	39.3	52.3
Asking students to review each other's work	Overall	16.6	37.5	33.4	12.5
	Americas	11.9	39.9	39.9	8.3
	Pacific	15.8	33.7	30.7	19.8
	Europe	19.8	37.3	30.2	12.7
Making connections with key mathematical concepts at lower or higher grade levels	Overall	4.5	18.2	43.0	34.3
	Americas	2.6	15.1	45.8	36.5
	Pacific	4.0	20.8	36.6	38.6
	Europe	5.9	19.2	43.3	31.6
Applying mathematical principles in real-world settings	Overall	0.7	10.5	34.2	54.7
	Americas	1.0	6.2	37.3	55.4
	Pacific	0.0	7.9	28.7	63.4
	Europe	0.7	14.0	34.1	51.3

		None	Minor	Moderate	Major
Giving tests and quizzes to find out what students have learned	Overall	4.0	17.1	40.3	38.6
	Americas	3.6	16.1	42.0	38.3
	Pacific	4.0	20.8	44.6	30.7
	Europe	4.2	16.6	37.8	41.4
Asking students to explain their thinking	Overall	0.7	5.8	28.4	65.1
	Americas	0.5	4.7	30.1	64.8
	Pacific	0.0	8.9	25.7	65.4
	Europe	1.0	5.5	28.3	65.3
Using correct mathematical terminology	Overall	0.8	3.0	26.6	69.6
	Americas	0.5	2.1	23.8	73.6
	Pacific	0.0	5.9	24.8	69.3
	Europe	1.3	2.6	28.9	67.2
Making connections between mathematics and other subjects	Overall	1.0	18.3	40.4	40.4
	Americas	1.0	13.5	44.0	41.5
	Pacific	0.0	20.8	40.6	38.6
	Europe	1.3	20.5	38.0	40.3
Having students use tools and technology in class to support their mathematics learning	Overall	1.0	12.5	40.4	46.2
	Americas	1.0	5.7	43.0	50.3
	Pacific	1.0	12.9	34.7	51.5
	Europe	1.0	16.6	40.6	41.9
Explaining mathematical principles	Overall	2.7	15.1	37.9	44.4
	Americas	1.6	14.0	35.2	49.2
	Pacific	0.0	16.8	36.6	46.5
	Europe	4.2	15.3	39.9	40.6
Using informal questions to assess student understanding	Overall	0.3	5.5	35.2	59.0
	Americas	0.5	5.2	35.2	59.1
	Pacific	0.0	4.0	36.6	59.4
	Europe	0.3	6.2	34.7	58.8
Showing students how to apply mathematical procedures to solve problems	Overall	0.8	9.0	43.2	47.0
	Americas	0.5	4.2	24.4	71.0
	Pacific	0.0	5.9	38.6	55.5
	Europe	1.6	5.2	32.1	61.0

NOTES: All data in this appendix are given as percentages. Percentages may not add up to 100 due to rounding.

5. In 2014–2015, how frequently do you do each of the following activities to adjust instruction to meet the needs of individual students in your mathematics classes (including students with disabilities and English learners)? (*N* = 587–591)

		Never	1–2 per Year	1–2 per Month	1–2 per Week	1+ per Lesson
Review assessment results to identify individual students who need supplemental instruction	Overall	0.0	3.4	27.6	42.0	27.1
	Americas	0.0	2.1	24.7	51.1	22.1
	Pacific	0.0	3.0	31.3	37.4	28.3
	Europe	0.0	4.3	28.2	37.8	29.8
Plan different assignments or lessons for individuals or groups of students based on their performance	Overall	0.0	2.7	14.3	50.3	32.8
	Americas	1.0	5.1	22.2	44.4	27.3
	Pacific	2.3	4.7	20.3	45.0	27.7
	Europe	0.0	2.7	14.3	50.3	32.8
Have a teacher aide provide help to individuals or groups of students during classroom instruction	Overall	35.8	2.4	5.8	30.3	25.7
	Americas	36.7	2.7	3.2	31.9	25.5
	Pacific	39.8	6.1	9.2	28.6	16.3
	Europe	33.9	1.0	6.3	29.9	28.9
Provide help to individual students outside of class time	Overall	20.8	6.8	16.8	39.0	16.7
	Americas	24.9	7.9	16.9	36.0	14.3
	Pacific	16.3	5.1	18.4	43.9	16.3
	Europe	19.6	6.6	16.3	39.2	18.3
Work one-on-one with a student during class	Overall	1.1	1.1	2.7	31.8	63.5
	Americas	0.0	1.0	2.0	32.3	64.7
	Pacific	1.3	0.7	5.0	35.1	58.0
	Europe	1.1	1.1	2.7	31.8	63.5
Translate lessons or assignments into other languages	Overall	79.8	5.9	4.2	5.9	4.1
	Americas	80.5	3.7	3.7	5.8	6.3
	Pacific	75.5	9.2	9.2	5.1	1.0
	Europe	80.8	6.3	3.0	6.3	3.6
Review student Individual Education Plan to understand student strengths and challenges	Overall	6.3	30.5	32.7	18.5	12.0
	Americas	4.7	26.3	31.6	19.5	17.9
	Pacific	6.1	30.6	40.8	14.3	8.2
	Europe	7.3	33.1	30.8	19.2	9.6
Talk with parent or guardian about how to support student learning	Overall	0.9	16.9	50.1	24.9	7.3
	Americas	0.5	13.2	47.9	28.4	10.0
	Pacific	0.0	16.2	49.5	26.3	8.1
	Europe	1.3	19.5	51.7	22.2	5.3

NOTES: All data in this appendix are given as percentages. Percentages may not add up to 100 due to rounding.

6. Did you receive any professional development this school year (including the summer of 2014)? (N = 588)

	Overall
Yes	66.3
No	33.7

NOTES: All data in this appendix are given as percentages. Percentages may not add up to 100 due to rounding.

7. How much has your professional development in mathematics focused on the following topics this school year (2014–2015, including summer 2014)? (N = 383–386)

		None	Minor	Moderate	Major
Content of the mathematics standards	Overall	27.2	27.7	23.3	21.8
	Americas	22.5	30.5	25.8	21.2
	Pacific	29.3	27.6	24.1	19.0
	Europe	30.5	25.4	20.9	23.2
Instructional strategies for teaching students to meet the mathematics	Overall	23.6	23.4	26.8	26.2
	Americas	19.2	25.8	25.8	29.1
	Pacific	36.2	22.4	22.4	19.0
	Europe	23.3	21.6	29.0	26.1
Development of classroom assessments	Overall	34.6	26.8	25.0	13.5
	Americas	30.5	29.8	25.8	13.9
	Pacific	45.6	19.3	21.1	14.0
	Europe	34.7	26.7	25.6	13.1
Use of assessment data to inform instruction	Overall	19.8	21.7	28.7	29.8
	Americas	16.6	19.2	31.1	33.1
	Pacific	28.1	22.8	28.1	21.1
	Europe	20.0	23.4	26.9	29.7
Use of student work examples to inform instruction	Overall	23.8	28.2	25.3	22.7
	Americas	20.0	26.0	30.7	23.3
	Pacific	29.8	28.1	19.3	22.8
	Europe	25.0	30.1	22.7	22.2
Using technology to support student mathematics learning	Overall	25.3	28.5	27.2	19.1
	Americas	23.2	23.2	31.1	22.5
	Pacific	31.6	36.8	21.1	10.5
	Europe	25.1	30.3	25.7	18.9

Differentiation of instruction for students at different achievement levels	Overall	30.0	21.1	27.3	21.6
	Americas	23.8	19.2	31.8	25.2
	Pacific	35.1	28.1	21.1	15.8
	Europe	33.5	20.5	25.6	20.5
Differentiation of instruction for ELs	Overall	55.5	23.4	11.5	9.6
	Americas	57.0	20.5	12.6	9.9
	Pacific	52.6	26.3	12.3	8.8
	Europe	55.1	25.0	10.2	9.7

		None	Minor	Moderate	Major
Differentiation of instruction for students with disabilities	Overall	47.3	24.5	16.5	11.8
	Americas	42.7	23.3	19.3	14.7
	Pacific	54.4	28.1	5.3	12.3
	Europe	48.9	24.4	17.6	9.1

NOTES: All data in this appendix are given as percentages. Percentages may not add up to 100 due to rounding.

8. How much do you need additional professional development on the following knowledge and skills to improve your mathematics instruction? (*N* = 573–574)

		None	Small	Moderate	High
Content of the mathematics standards	Overall	39.4	32.1	17.8	10.7
	Americas	42.3	34.6	13.2	9.9
	Pacific	33.3	29.2	21.9	15.6
	Europe	39.7	31.5	19.3	9.5
Instructional strategies for teaching students to meet the mathematics	Overall	22.7	34.8	27.4	15.2
	Americas	23.5	41.0	25.1	10.4
	Pacific	20.8	25.0	28.1	26.0
	Europe	22.7	34.2	28.5	14.6
Development of classroom assessments	Overall	24.4	32.1	26.0	17.6
	Americas	30.6	32.2	22.4	14.8
	Pacific	22.9	21.9	28.1	27.1
	Europe	21.0	35.3	27.5	16.3
Use of assessment data to inform instruction	Overall	33.5	35.4	18.1	13.1
	Americas	42.1	35.0	14.2	8.7
	Pacific	32.3	22.9	21.9	22.9
	Europe	28.5	39.7	19.3	12.5

		None	Small	Moderate	High
Use of student work examples to inform instruction	Overall	34.9	33.9	21.1	10.1
	Americas	44.3	35.5	13.1	7.1
	Pacific	33.7	23.2	24.2	19.0
	Europe	29.5	36.3	25.1	9.2
Using technology to support student mathematics learning	Overall	17.4	29.3	32.6	20.7
	Americas	18.6	32.2	31.7	17.5
	Pacific	15.6	29.2	25.0	30.2
	Europe	17.3	27.5	35.6	19.7
Differentiation of instruction for students at different achievement levels	Overall	21.8	29.1	29.1	20.0
	Americas	24.6	30.1	31.2	14.2
	Pacific	17.7	24.0	25.0	33.3
	Europe	21.4	30.2	29.2	19.3
Differentiation of instruction for ELs	Overall	24.4	34.3	24.2	17.1
	Americas	18.6	38.3	25.7	17.5
	Pacific	20.8	22.9	28.1	28.1
	Europe	29.2	35.6	22.0	13.2
		None	**Small**	**Moderate**	**High**
Differentiation of instruction for students with disabilities	Overall	22.9	35.8	25.0	16.4
	Americas	21.3	38.8	26.2	13.7
	Pacific	17.7	34.4	20.8	27.1
	Europe	25.5	34.4	25.5	14.6

NOTES: All data in this appendix are given as percentages. Percentages may not add up to 100 due to rounding.

9. How much do you need additional professional development on the following mathematics topics to improve your mathematics instruction? (N = 560–561)

		None	Small	Moderate	High
Developing students' conceptual understanding of mathematics	Overall	23.7	30.8	28.9	16.6
	Americas	27.0	37.1	24.7	11.2
	Pacific	23.4	19.2	29.8	27.7
	Europe	21.8	30.8	31.1	16.3
Helping students master basic mathematical skills and procedures	Overall	35.5	34.5	20.4	9.6
	Americas	40.5	34.8	16.9	7.9
	Pacific	35.1	26.6	23.4	14.9
	Europe	32.6	36.8	21.5	9.0

		None	Small	Moderate	High
Helping students develop fluency with mathematical skills and procedures	Overall	25.1	37.3	24.1	13.6
	Americas	30.3	42.7	18.0	9.0
	Pacific	20.2	26.6	33.0	20.2
	Europe	23.5	37.4	24.9	14.2
Helping students use tools and technology in class to support their mathematics learning	Overall	20.2	31.4	32.5	15.9
	Americas	23.2	37.3	26.6	13.0
	Pacific	20.2	24.5	34.0	21.3
	Europe	18.3	30.1	35.6	15.9
Making connections between key mathematics topics and concepts within my grade level(s) or course(s)	Overall	30.0	37.1	23.2	9.6
	Americas	37.1	37.1	19.1	6.7
	Pacific	29.8	28.7	27.7	13.8
	Europe	25.7	39.9	24.3	10.1
Making connections between key mathematics topics and concepts across grade levels	Overall	25.0	35.1	27.3	12.7
	Americas	29.2	36.0	24.2	10.7
	Pacific	23.4	24.5	34.0	18.1
	Europe	22.8	38.1	27.0	12.1
Helping students make sense of problems and persevere in solving them	Overall	21.8	28.9	31.0	18.4
	Americas	25.8	31.5	25.3	17.4
	Pacific	23.4	23.4	30.9	22.3
	Europe	18.7	29.1	34.6	17.7
Helping students use mathematical language and symbols appropriately	Overall	32.1	35.5	22.1	10.3
	Americas	39.3	33.2	19.7	7.9
	Pacific	26.6	31.9	26.6	14.9
	Europe	29.4	38.1	22.2	10.4

		None	Small	Moderate	High
Helping students construct viable arguments and critique the reasoning of others	Overall	17.5	28.0	33.5	21.0
	Americas	16.9	30.9	34.8	17.4
	Pacific	13.8	26.6	28.7	30.9
	Europe	19.0	26.6	34.3	20.1
Helping students apply mathematics to solve problems in real-world contexts	Overall	23.9	28.7	28.9	18.5
	Americas	30.3	32.6	21.9	15.2
	Pacific	23.4	20.2	28.7	27.7
	Europe	20.1	29.1	33.2	17.7

Helping students look for and make use of structure (e.g., patterns in numbers, shapes, or algorithms)	Overall	27.5	36.5	23.7	12.3
	Americas	30.9	37.6	21.4	10.1
	Pacific	28.7	27.7	28.7	14.9
	Europe	24.9	38.8	23.5	12.8
Understanding the difference between mathematics concepts addressed by the CCR standards for mathematics and those addressed by the DoDEA mathematics standards	Overall	9.8	16.6	26.1	47.5
	Americas	7.9	18.5	30.3	43.3
	Pacific	13.8	12.8	26.6	46.8
	Europe	9.7	16.7	23.3	50.4
Understanding which mathematics concepts are being prioritized in the CCR standards for mathematics	Overall	9.3	15.0	23.7	52.1
	Americas	9.0	15.2	28.1	47.8
	Pacific	10.6	11.7	22.3	55.3
	Europe	9.0	15.9	21.5	53.6

NOTE: Percentages have been rounded.

References

Achieve and Education First, *A Strong State Role in Common Core State Standards Implementation: Rubric and Self-Assessment Tool*, March 2012. As of July 29, 2015:
http://www.achieve.org/common-core-state-standards-implementation-rubric-and-self-assessment-tool

Achieve, College Summit, National Association of Secondary School Principals, and National Association of Elementary School Principals, *Implementing the Common Core State Standards: The Role of the Secondary School Leader*, February 2013. As of August 12, 2015:
http://www.achieve.org/publications/implementing-common-core-state-standards-role-secondary-school-leader-action-brief

Afterschool Alliance, "Partnerships with STEM-Rich Institutions," November 2013. As of August 12, 2015:
http://www.afterschoolalliance.org/issue_briefs/issue_STEM_61.pdf

Alberti, Sandra, "Making the Shifts," *Educational Leadership*, Vol. 70, No. 4, December 2012–January 2013, pp. 24–27.

ASCD—See Association for Supervision and Curriculum Development.

Aspen Institute Education and Society Program, Education First, Insight Education Group, Student Achievement Partners, and Targeted Leadership Consulting, *Implementation of the Common Core State Standards: A Transition Guide for School-Level Leaders*, Washington, D.C.: Aspen Institute, 2013. As of March 1, 2016:
http://www.aspendrl.org/portal/browse/DocumentDetail?documentId=1882&download

Association for Supervision and Curriculum Development, *Developing Performance Assessments: Facilitator's Guide*, Alexandria, Va., 1996. As of March 1, 2016:
http://shop.ascd.org/ASCD/pdf/pdi/DevelopingPerformance.pdf

———, *Fulfilling the Promise of the Common Core State Standards: Moving from Adoption to Implementation and Sustainability*, Alexandria, Va., 2012.

Balfanz, Robert, Douglas J. Mac Iver, and Vaughan Byrnes, "The Implementation and Impact of Evidence-Based Mathematics Reforms in High-Poverty Middle Schools: A Multi-Site, Multi-Year Study," *Journal for Research in Mathematics Education*, Vol. 37, No. 1, 2006, pp. 33–64.

Ball, Deborah L., and David K. Cohen, "Reform by the Book: What Is: Or Might Be: The Role of Curriculum Materials in Teacher Learning and Instructional Reform?" *Educational Researcher*, Vol. 25, No. 9, 1996, pp. 6–14.

Ball, Deborah Loewenberg, and Francesca M. Forzani, "Building a Common Core for Learning to Teach: And Connecting Professional Learning to Practice," *American Educator*, Vol. 35, No. 2, 2011, pp. 17–39.

Ball, Deborah Loewenberg, Heather C. Hill, and Hyman Bass, "Knowing Mathematics for Teaching: Who Knows Mathematics Well Enough to Teach Third Grade, and How Can We Decide?" *American Educator*, Vol. 29, No. 1, 2005, pp. 14–17, 20–22, 43–46.

Ball, Deborah Loewenberg, Mark Hoover Thames, and Geoffrey Phelps, "Content Knowledge for Teaching: What Makes it Special?" *Journal of Teacher Education*, Vol. 59, No. 5, November/December 2008, pp. 389–407.

Barron, Brigid, Susie Wise, and Caitlin K. Martin, "Creating Within and Across Life Spaces: The Role of a Computer Clubhouse in a Child's Learning Ecology," in Bronwyn Bevan, Philip Bell, Reed Stevens, and Aria Razfar, eds., *LOST Opportunities: Learning in Out-of-School Time*, Dordrecht, Netherlands: Springer, 2012, pp. 99–118.

Bevan, Bronwyn, and Vera Michalchik, "Where It Gets Interesting: Competing Models of STEM Learning After School," *Afterschool Matters*, Spring 2013, pp. 1–8.

Black, Paul, and Dylan Wiliam, "Inside the Black Box: Raising Standards Through Classroom Assessment," November 11, 1998. As of March 1, 2016: http://www.spd.dcu.ie/site/teaching_today/documents/raisingstandardsthroughclassroomassessment.pdf

Boaler, Jo, *Experiencing School Mathematics: Traditional and Reform Approaches to Teaching and Their Impact on Student Learning*, revised and expanded edition, Mahwah, N.J.: Lawrence Erlbaum Associations, Inc., Publishers, 2002.

Boaler, Jo, and Megan Staples, "Creating Mathematical Futures Through an Equitable Teaching Approach: The Case of Railside School," *Teachers College Record*, Vol. 110, No. 3, March 2008, pp. 608–645.

Borko, Hilda, "Professional Development and Teacher Learning: Mapping the Terrain," *Educational Researcher*, Vol. 33, No. 8, November 2004, pp. 3–15.

Bossert, Steven T., David C. Dwyer, Brian Rowan, and Ginny V. Lee, "The Instructional Management Role of the Principal," *Education Administration Quarterly*, Vol. 18, No. 3, Summer 1982, pp. 34–64.

Boston, Melissa D., and Margaret S. Smith, "Transforming Secondary Mathematics Teaching: Increasing the Cognitive Demands of Instructional Tasks Used in Teachers' Classrooms," *Journal for Research in Mathematics Education*, Vol. 40, No. 2, March 2009, pp. 119–156.

Bransford, John D., Ann L. Brown, and Rodney R. Cocking, eds., *How People Learn: Brain, Mind, Experience, and School*, Washington, D.C.: National Academies Press, 1999.

Brookhart, Susan M., "Educational Assessment Knowledge and Skills for Teachers," *Educational Measurement: Issues and Practice,* Vol. 30, No. 1, Spring 2011, pp. 3–12.

Bulgren, Janis A., B. Keith Lenz, Donald D. Deshler, and Jean B. Schumaker, *The Content Enhancement Series: The Concept Comparison Routine*, Lawrence, Kan.: Edge Enterprises, 1995.

Carpenter, Thomas P., Elizabeth Fennema, Penelope L. Peterson, Chi-Pang Chiang, and Megan Loef, "Using Knowledge of Children's Mathematics Thinking in Classroom Teaching: An Experimental Study," *American Educational Research Journal*, Vol. 26, No. 4, Winter 1989, pp. 499–531.

CCSSO—See Council of Chief State School Officers.

Charalambous, Charalambos Y., "Mathematical Knowledge for Teaching and Task Unfolding: An Exploratory Study," *Elementary School Journal*, Vol. 110, No. 3, March 2010, pp. 247–278.

Clark, Charles S., "Director of Pentagon's Far-Flung School System Counts His Blessings," *Government Executive*, June 4, 2015. As of November 20, 2015: http://www.govexec.com/management/2015/06/ director-pentagons-far-flung-school-system-counts-his-blessings/114523/

Cobb, Paul, Terry Wood, Erna Yackel, John Nicholls, Grayson Wheatley, Beatriz Trigatti, and Marcella Perlwitz, "Assessment of a Problem-Centered Second-Grade Mathematics Project," *Journal for Research in Mathematics Education*, Vol. 22, No. 1, January 1991, pp. 3–29.

Coburn, Cynthia Ellen, *Making Sense of Reading: Logics of Reading in the Institutional Environment and the Classroom*, dissertation, Palo Alto, Calif.: Stanford University, 2001.

Cole, Camille, *Connecting Students to STEM Careers: Social Networking Strategies*, Arlington, Va.: International Society for Technology in Education, 2011.

Coleman, James S., "The Concept of Equality of Educational Opportunity," *Harvard Educational Review*, Vol. 38, No. 1, Spring 1968, pp. 7–22.

Common Core State Standards Initiative, "About the Standards," undated (a). As of May 5, 2016: http://www.corestandards.org/about-the-standards

———, "Mathematical Practice," undated (b). As of May 5, 2016:
http://www.corestandards.org/Math/

———, "Standards for Mathematical Practice," undated (c). As of May 5, 2016:
http://www.corestandards.org/Math/Practice/

"Conclusions and Recommendations," in Jeremy Kilpatrick, Jane Swafford, and Bradford Findell, eds., *Adding It Up: Helping Children Learn Mathematics*, Washington, D.C.: National Academies Press, 2001, pp. 407–432.

Confrey, Jere, and Erin Krupa, "Curriculum Design, Development and Implementation in an Era of Common Core State Standards: Summary Report of a Conference," Arlington, Va., August 1–3, 2010. As of July 29, 2015:
http://mathcurriculumcenter.org/conferences/ccss/SummaryReportCCSS

Corcoran, Tom, Siobhan McVay, and Kate Riordan, *Getting It Right: The MISE Approach to Professional Development, Philadelphia, Pa.: Consortium for Policy Research in Education*, Research Report, RR-055, 2003. As of March 1, 2016:
http://www.cpre.org/getting-it-right-mise-approach-professional-development

Correnti, Richard, "An Empirical Investigation of Professional Development Effects on Literacy Instruction Using Daily Logs," *Educational Evaluation and Policy Analysis*, Vol. 29, No. 4, December 2007, pp. 262–295.

Correnti, Richard, and Brian Rowan, "Opening Up the Black Box: Literacy Instruction in Schools Participating in Three Comprehensive School Reform Programs," *American Educational Research Journal*, Vol. 44, No. 2, June 2007, pp. 298–338.

Cullum, J., C. Hailey, D. Householder, C. Merrill, and J. Dorward, "Formative Evaluation of a Professional Development Program for High School Teachers Infusing Engineering Design into the Classroom," presentation at the meeting of the American Society for Engineering Education, Pittsburgh, Pa., 2008. As of August 12, 2015:
http://digitalcommons.usu.edu/ncete_present/22/

Darling-Hammond, Linda, *The Right to Learn: A Blueprint for Creating Schools That Work*, San Francisco, Calif.: Jossey-Bass, 1997.

Darling-Hammond, Linda, Joan Herman, James Pellegrino, Jamal Abedi, J. Lawrence Aber, Eva Baker, Randy Bennett, Edmund Gordon, Edward Haertel, Kenji Hakuta, Andrew Ho, Robert Lee Linn, P. David Pearson, James Popham, Lauren Resnick, Alan H. Schoenfeld, Richard Shavelson, Lorrie A. Shepard, Lee Shulman, and Claude M. Steele, *Criteria for High-Quality Assessment*, Stanford, Calif.: Stanford Center for Opportunity Policy in Education, Stanford University; Center for Research on Student Standards and Testing, University of California at Los Angeles; and Learning Sciences Research Institute, University of Illinois at Chicago, June 19, 2013. As of March 1, 2016:
https://edpolicy.stanford.edu/sites/default/files/publications/criteria-higher-quality-assessment_2.pdf

Datnow, Amanda, Geoffrey Borman, and Sam Stringfield, "School Reform Through a Highly Specified Curriculum: Implementation and Effects of the Core Knowledge Sequence," *Elementary School Journal*, Vol. 101, No. 2, November 2000, pp. 167–191.

Davis, Elizabeth A., and Joseph S. Krajcik, "Designing Educative Curriculum Materials to Promote Teacher Learning," *Educational Researcher*, Vol. 34, No. 3, April 2005, pp. 3–14.

Department of Defense Education Activity, "Community Strategic Plan, School Years 2013/14–2017/18," undated. As of May 5, 2016:
http://www.dodea.edu/CSP/

Desimone, Laura M., "Improving Impact Studies of Teachers' Professional Development: Toward Better Conceptualizations and Measures," *Educational Researcher*, Vol. 38, No. 3, 2009, pp. 181–199.

DoDEA—See Department of Defense Education Activity.

Doyle, Walter, "Academic Work," *Review of Educational Research*, Vol. 53, No. 2, Summer 1983, pp. 159–199.

Eccles, Jacquelynne S., "Understanding Women's Educational and Occupational Choices: Applying the Eccles et al. Model of Achievement-Related Choices," *Psychology of Women Quarterly*, Vol. 18, 1994, pp. 585–609.

Education First and Achieve, "A Strong State Role in Common Core State Standards Implementation: Rubric and Self-Assessment Tool," March 2012. As of August 12, 2015:
http://www.achieve.org/files/Achieve-CCSSrubricandstatetoolFINAL.pdf

Elliott, Rebekah, Elham Kazemi, Kristin Lesseig, Judith Mumme, Cathy Carroll, and Megan Kelley-Petersen, "Conceptualizing the Work of Leading Mathematical Tasks in Professional Development," *Journal of Teacher Education*, Vol. 60, No. 4, 2009, pp. 364–379.

Elmore, Richard F., *Bridging the Gap Between Standards and Achievement: The Imperative for Professional Development in Education*, Washington, D.C.: Albert Shanker Institute, 2002. As of March 1, 2016:
http://www.shankerinstitute.org/sites/shanker/files/Bridging_Gap.pdf

Equity Assistance Centers, "How the Common Core Must Ensure Equity by Fully Preparing Every Student for Postsecondary Success: Recommendations from the Regional Equity Assistance Centers on Implementation of the Common Core State Standards," 2013. As of May 5, 2016:
http://educationnorthwest.org/sites/default/files/resources/EquityCommonCore_120913.pdf

Evans, Paula M., and Nancy Mohr, "Professional Development for Principals: Seven Core Beliefs," *Phi Delta Kappan*, Vol. 80, No. 7, March 1999, pp. 530–533.

Fullan, Michael, *The New Meaning of Educational Change*, 2nd ed., New York: Teachers College Press, Columbia University, 1991.

Garet, Michael S., Andrew C. Porter, Laura Desimone, Beatrice F. Birman, and Kwang Suk Yoon, "What Makes Professional Development Effective? Results from a National Sample of Teachers," *American Educational Research Journal*, Vol. 38, No. 4, December 2001, pp. 915–945.

"Gearing Up for the Common Core State Standards in Mathematics: Five Initial Domains for Professional Development in Grades K–8," May 7, 2011. As of March 3, 2016:
https://commoncoretools.files.wordpress.com/2011/05/2011_05_07_gearing_up1.pdf

Goldenberg, Claude, and Ronald Gallimore, "Changing Teaching Takes More Than a One-Shot Workshop," *Educational Leadership*, Vol. 49, No. 3, November 1991, pp. 69–72.

Goldman, Susan, and Ted S. Hasselbring, "Achieving Meaningful Mathematics Literacy for Students with Learning Disabilities," *Journal of Learning Disabilities*, Vol. 30, No. 2, March–April 1997, pp. 198–208.

Goldring, Ellen B., Courtney Preston, and Jason Huff, "Conceptualizing and Evaluating Professional Development for School Leaders," *Planning and Changing*, Vol. 43, Nos. 3/4, 2012, pp. 223–242.

Grissom, Jason A., Susanna Loeb, and Benjamin Master, "Effective Instructional Time Use for School Leaders: Longitudinal Evidence from Observations of Principals," *Educational Researcher*, Vol. 42, No. 8, November 2013, pp. 433–444.

Guskey, T. R., and K. S. Yoon, "What Works in Professional Development," *Phi Delta Kappan*, Vol. 90, No. 7, 2009, pp. 495–500.

Hakuta, Kenji, and María Santos, *Understanding Language: Challenges and Opportunities for Language Learning in the Context of Common Core State Standards and Next Generation Science Standards*, conference overview paper, Stanford, Calif.: Stanford University, 2012. As of March 1, 2016:
http://ell.stanford.edu/sites/default/files/Conference%20Summary_0.pdf

Halpern, Diane F., Joshua Aronson, Nona Reimer, Sandra Simpkins, Jon R. Star, and Kathryn Wentzel, *Encouraging Girls in Math and Science: IES Practice Guide*, Washington, D.C.: National Center for Education Research, Institute of Education Sciences, U.S. Department of Education, NCER 2007–2003, September 2007. As of March 1, 2016:
http://ies.ed.gov/ncee/wwc/pdf/practice_guides/20072003.pdf

Hamilton, Laura S., Brian M. Stecher, and Kun Yuan, *Standards Based Reform in the United States: History, Research, and Future Directions*, Santa Monica, Calif.: RAND Corporation, RP-1384, 2008. As of March 1, 2016:
http://www.rand.org/pubs/reprints/RP1384.html

Hattie, John, and Helen Timperley, "The Power of Feedback," *Review of Educational Research*, Vol. 77, No. 1, March 2007, pp. 81–112.

Hiebert, James, "What Research Says About the NCTM Standards," in Jeremy Kilpatrick, W. Gary Martin, and Deborah Schifter, eds., *A Research Companion to Principles and Standards for School Mathematics*, Reston, Va.: National Council of Teachers of Mathematics, 2003, pp. 5–23.

———, Conceptual and Procedural Knowledge: The Case of Mathematics, New York: Routledge, 2013.

Hiebert, James, and Diana Wearne, "Instructional Tasks, Classroom Discourse, and Students' Learning in Second-Grade Arithmetic," *American Educational Research Journal*, Vol. 30, No. 2, June 1993, pp. 393–425.

Hill, Heather C., Laura Kapitula, and Kristin Umland, "A Validity Argument Approach to Evaluating Teacher Value-Added Scores," *American Educational Research Journal*, Vol. 48, No. 3, June 2011, pp. 794–831.

Hill, Heather C., Deborah Loewenberg Ball, and Stephen G. Schilling, "Unpacking Pedagogical Content Knowledge: Conceptualizing and Measuring Teachers' Topic-Specific Knowledge of Students," *Journal for Research in Mathematics Education*, Vol. 39, No. 4, 2008, pp. 372–400.

Hill, Heather C., Brian Rowan, and Deborah Loewenberg Ball, "Effects of Teachers' Mathematical Knowledge for Teaching on Student Achievement," *American Educational Research Journal*, Vol. 42, No. 2, Summer 2005, pp. 371–406.

Hynes, Morgan M., and Angel Dos Santos, "Effective Teacher Professional Development: Middle School Engineering Content," *International Journal of Engineering Education*, Vol. 23, No. 1, February 2007, pp. 24–29.

Johnston, Tad, Robin Bzura, Steve Leinwand, Beth Ratway, Tori Cirks, and Asta Svedkauskaite, "DoDEA Mathematics Standards Gap Analysis: Alignment of DoDEA's College and Career Ready Standards for Mathematics with DoDEA Academic Standards," undated. As of March 2, 2016: https://content.dodea.edu/VS/pd/ccrsm_training/docs/may/handouts/ho_9_gap_analysis_sample_packet_04-13-2015.pdf

Kane, Michael T., "Terminology, Emphasis, and Utility in Validation," *Educational Researcher*, Vol. 37, No. 2, March 2008, pp. 76–82.

Kanold, Timothy D., and Matthew R. Larson, *Common Core Mathematics in a PLC at Work: Leader's Guide*, Bloomington, Ind.: Solution Tree Press, 2012.

Kelley, Carolyn, and Kent D. Peterson, "Principal Inservice Programs: A Portrait of Diversity and Promise," in Marc S. Tucker and Judy B. Codding, eds., *The Principal Challenge: Leading and Managing Schools in an Era of Accountability*, San Francisco, Calif.: Jossey-Bass, 2002, pp. 313–346.

Kendall, John S., *Understanding Common Core State Standards*, Alexandria, Va.: Association for Supervision and Curriculum Development, 2011.

Kepner, Henry S., and DeAnn Huinker, "Assessing Students' Mathematical Proficiencies on the Common Core," *Journal of Mathematics Education at Teachers College*, Vol. 3, No. 1, Spring–Summer 2012, pp. 26–32.

Khisty, Lena Licón, and Craig Joseph Willey, "After-School: An Innovative Model to Better Understand the Mathematics Learning of Latinas/os," in Bronwyn Bevan, Philip Bell, Reed Stevens, and Aria Razfar, eds., *LOST Opportunities: Learning in Out-of-School Time*, Dordrecht, Netherlands: Springer, 2012, pp. 233–250.

Kilpatrick, Jeremy, Jane Swafford, and Bradford Findell, eds., "Conclusions and Recommendations," in *Adding It Up: Helping Children Learn Mathematics*, Washington, D.C.: National Academies Press, 2001, pp. 407–432.

Knapp, Michael S., "Between Systemic Reforms and the Mathematics and Science Classroom: The Dynamics of Innovation, Implementation, and Professional Learning," *Review of Educational Research*, Vol. 67, No. 2, Summer 1997, pp. 227–266.

Koch, Melissa, Annie Georges, Torie Gorges, and Reina Fujii, "Engaging Youth with STEM Professionals in Afterschool Programs," *Meridian*, Vol. 13, No. 1, 2010, pp. 1–15. As of August 12, 2015: http://www.ncsu.edu/meridian/winter2010/koch/

Krajcik, Joseph, Katherine L. McNeill, and Brian J. Reiser, "Learning-Goals-Driven Design Model: Developing Curriculum Materials That Align with National Standards and Incorporate Project-Based Pedagogy," Science Education, Vol. 92, No. 1, 2008, pp. 1–32.

Krishnamurthi, Anita, Ron Ottinger, and Tessie Topol, "STEM Learning in Afterschool and Summer Programming: An Essential Strategy for STEM Education Reform," undated. As of August 12, 2015:
http://www.expandinglearning.org/expandingminds/article/
stem-learning-afterschool-and-summer-programming-essential-strategy-stem

Krupa, Erin, *A Summary Report from the Conference "Moving Forward Together: Curriculum and Assessment and the Common Core State Standards for Mathematics,"* Arlington, Va., April 29–May 1, 2011.

Lampert, Magdalene, "When the Problem Is Not the Question and the Solution Is Not the Answer: Mathematical Knowing and Teaching," *American Educational Research Journal*, Vol. 27, No. 1, Spring 1990, pp. 29–63.

Lappan, Glenda, and Elizabeth Phillips, "Teaching and Learning in the Connected Mathematics Project," in Larry Leutzinger, ed., *Mathematics in the Middle*, Reston, Va,: National Council of Teachers of Mathematics, 1998, pp. 83–92.

Larson, Matthew R., *Administrator's Guide: Interpreting the Common Core State Standards to Improve Mathematics Education*, Reston, Va: National Council of Teachers of Mathematics, 2011.

Lawrence, Marlyn, Helen Santiago, Kathy Zamora, Albert Bertani, and Rob Bocchino, "Practice as Performance," *Principal Leadership*, Vol. 9, No. 4, December 2008, pp. 32–38.

Lazer, Stephen, John Mazzeo, Jon S. Twing, Walter D. Way, Wayne Camara, and Kevin Sweeney, *Thoughts on an Assessment of Common Core Standards*, Princeton, N.J.: Educational Testing Service, 2010.

Leithwood, Kenneth, Karen Seashore Louis, Stephen Anderson, and Kyla Wahlstrom, "Review of Research: How Leadership Influences Student Learning," 2004. As of August 12, 2015:
http://www.wallacefoundation.org/knowledge-center/school-leadership/key-research/Documents/
How-Leadership-Influences-Student-Learning.pdf

Liston, Carrie, Karen Peterson, and Vicky Ragan, "Guide to Promising Practices in Informal Information Technology Education for Girls," Boulder, Colo.: National Center for Women and Information Technology and Girl Scouts, 2007.

Loveless, Tom, "The Common Core Initiative: What Are the Chances of Success?" *Educational Leadership*, Vol. 70, No. 4, December 2012–January 2013, pp. 60–63.

Marrongelle, Karen, Paola Sztajn, and Margaret Smith, "Scaling Up Professional Development in an Era of Common State Standards," Journal of Teacher Education, Vol. 64, No. 3, 2013, pp. 202–211.

McIntosh, Margaret E., "Formative Assessment in Mathematics," *Clearing House: A Journal of Educational Strategies, Issues, and Ideas*, Vol. 71, No. 2, 1997, pp. 92–96.

Miller, Susan P., and Pamela J. Hudson, "Using Evidence-Based Practices to Build Mathematics Competence Related to Conceptual, Procedural, and Declarative Knowledge," *Learning Disabilities Research and Practice*, Vol. 22, No. 1, February 2007, pp. 47–57.

Moschkovich, Judit, "Mathematics, the Common Core, and Language: Recommendations for Mathematics Instruction for ELs Aligned with the Common Core," *Proceedings of the Thirty-Fourth Annual Meeting of the North American Chapter of the International Group for the Psychology of Mathematics Education*, Kalamazoo, Mich.: Western Michigan University, November 1–4, 2012.

Murphy, Joseph, "Equity as Student Opportunity to Learn," *Theory into Practice*, Vol. 27, No. 2, 1988, pp. 145–151.

NAESP—See National Association of Elementary School Principals.

National Association of Elementary School Principals, "Common Core Implementation Checklist for Principals," May 2012. As of August 12, 2015:
http://www.naesp.org/communicator-may-2012/common-core-implementation-checklist-principals

National Council of Teachers of Mathematics, *Principles and Standards for School Mathematics*, Reston, Va., 2000.

———, *Curriculum Focal Points for Prekindergarten Through Grade 8 Mathematics: A Quest for Coherence*, Reston, Va., 2006.

National Research Council, *Adding It Up: Helping Children Learn Mathematics*, Washington, D.C.: National Academies Press, 2001.

National Staff Development Council, *Learning to Lead, Leading to Learn: Improving School Quality Through Principal Professional Development*, Oxford, Ohio, 2000.

NCTM—See National Council of Teachers of Mathematics.

O'Day, Jennifer A., and Marshall S. Smith, "Systemic Reform and Educational Opportunity," in Susan H. Fuhrman, ed., *Designing Coherent Education Policy: Improving the System*, San Francisco: Jossey-Bass, 1993.

Organization for Economic Co-operation and Development, "Strong Performers and Successful Reformers in Education: Lessons from PISA for the United States," Paris, France, May 11, 2011. As of May 5, 2016: http://www.oecd-ilibrary.org/education/lessons-from-pisa-for-the-united-states_9789264096660-en

Peterson, Kent, "The Professional Development of Principals: Innovations and Opportunities," *Educational Administration Quarterly*, Vol. 38, No. 2, April 2002, pp. 213–232.

Polikoff, Morgan S., Andrew C. Porter, and John Smithson, "How Well Aligned Are State Assessments of Student Achievement with State Content Standards?" *American Educational Research Journal*, Vol. 48, No. 4, 2011, pp. 965–995.

Porter, Andrew, Jennifer McMaken, Jun Hwang, and Rui Yang, "Common Core Standards: The New U.S. Intended Curriculum," *Educational Researcher*, Vol. 40, No. 3, April 2011, pp. 103–116.

Powell, Sarah R., Lynn S. Fuchs, and Doug Fuchs, "Reaching the Mountaintop: Addressing the Common Core Standards in Mathematics for Students with Mathematics Difficulties," *Learning Disabilities Research and Practice*, Vol. 28, No. 1, 2013, pp. 38–48.

Public Law 101-476, Individuals with Disabilities Education Act, October 30, 1990. As of May 17, 2016: http://uscode.house.gov/statutes/pl/101/476.pdf

Public Law 107-110, An Act to Close the Achievement Gap with Accountability, Flexibility, and Choice, So That No Child Is Left Behind, January 8, 2002. As of May 17, 2016: https://www.gpo.gov/fdsys/pkg/PLAW-107publ110/content-detail.html

RAND Education, "The American Teacher Panel and the American School Leader Panel," undated. As of May 5, 2016: http://www.rand.org/education/projects/atp-aslp.html

Richmond, Emily, "On Military Bases, Common Core by Another Name," Hechinger Report, March 6, 2015. As of December 6, 2015: http://www.usnews.com/news/articles/2015/03/06/schools-on-military-bases-opt-for-common-core-by-another-name

Riddle, Mel, "The Principal's Role in Implementing the Common Core State Standards: Ten Keys to Success," *High School Matters Blog*, 2012. As of August 12, 2015: http://www.ccrscenter.org/products-resources/blog/principal%E2%80%99s-role-implementing-common-core-state-standards-ten-keys-success

Sadler, D. Royce, "Formative Assessment and the Design of Instructional Systems," *Instructional Science*, Vol. 18, No. 2, 1989, pp. 119–144.

Sanders, Mark, "STEM, STEM Education, STEMmania," *Technology Teacher*, Vol. 68, No. 4, December 2008–January 2009, pp. 20–26.

Schmidt, William H., and Richard T. Houang, "Curricular Coherence and the Common Core State Standards for Mathematics," *Educational Researcher*, Vol. 41, No. 8, November 2012, pp. 294–308.

Schmidt, William H., Curtis C. McKnight, Richard T. Houang, Hsing Chi Wang, David E. Wiley, Leland S. Cogan, and Richard G. Wolfe, *Why Schools Matter: A Cross-National Comparison of Curriculum and Learning*, San Francisco: Jossey-Bass, 2001.

Schmidt, William H., Curtis C. McKnight, and Senta A. Raizen, *A Splintered Vision: An Investigation of U.S. Mathematics and Science Education*, New York: Kluwer Academic Publishers, 1997.

Schmidt, William H., Hsing Chi Wang, and Curtis C. McKnight, "Curriculum Coherence: An Examination of U.S. Mathematics and Science Content Standards from an International Perspective," *Journal of Curriculum Studies*, Vol. 37, No. 5, 2005, pp. 525–559.

Schoen, Harold L., Kristen J. Cebulla, Kelly F. Finn, and Cos Fi, "Teacher Variables That Relate to Student Achievement When Using a Standards-Based Curriculum," *Journal for Research in Mathematics Education*, Vol. 34, No. 3, May 2003, pp. 228–259.

Shaw, Jean M., "Manipulatives Enhance the Learning of Mathematics," *Houghton Mifflin Mathematics*, 2002. As of July 29, 2015:
http://www.eduplace.com/state/pdf/author/shaw.pdf

Simon, Martin A., and Deborah Schifter, "Towards a Constructivist Perspective: An Intervention Study of Mathematics Teacher Development," *Educational Studies in Mathematics*, Vol. 22, No. 4, August 1991, pp. 309–331.

Star, Jon R., "Reconceptualizing Procedural Knowledge," *Journal for Research in Mathematics Education*, Vol. 36, No. 5, November 2005, pp. 404–411.

Stein, Mary K., Jo Boaler, and Edward A. Silver, "Teaching Mathematics Through Problem Solving: Research Perspectives," in Harold L. Schoen, ed., *Teaching Mathematics Through Problem Solving, Grades 6–12*, Reston, Va.: National Council of Teachers of Mathematics, 2003, pp. 245–256.

Stein, Mary K., Barbara W. Grover, and Marjorie Henningsen, "Building Student Capacity for Mathematical Thinking and Reasoning: An Analysis of Tasks Used in Reform Classroom," *American Educational Research Journal*, Vol. 33, No. 2, Summer 1996, pp. 455–488.

Stein, Mary K., and Julia H. Kaufman, "Selecting and Supporting the Use of Mathematics Curricula at Scale," *American Educational Research Journal*, Vol. 43, No. 3, September 2010, pp. 663–693.

Supovitz, Jonathan A., and Herbert M. Turner, "The Effects of Professional Development on Science Teaching Practices and Classroom Culture," *Journal of Research in Science Teaching*, Vol. 37, No. 9, 2000, pp. 963–980.

Tamayo, Joaquin R., Jr., "Assessment 2.0: 'Next-Generation' Comprehensive Assessment Systems, an Analysis of Proposals by the Partnership for the Assessment of Readiness for College and Careers and SMARTER Balanced Assessment Consortium," Washington, D.C.: Aspen Institute Education and Society Program, 2010.

TNTP, *The Mirage: Confronting the Hard Truth About Our Quest for Teacher Development*, 2015. As of March 1, 2016:
http://tntp.org/assets/documents/TNTP-Mirage_2015.pdf

U.S. Department of Education, "Continuing to Expose and Close Achievement Gaps," undated (a). As of August 12, 2015:
https://www2.ed.gov/policy/elsec/guid/esea-flexibility/resources/close-achievement-gaps.pdf

———, "No Child Left Behind: Elementary and Secondary Education Act (ESEA)," undated (b). As of May 25, 2016:
http://www2.ed.gov/nclb/landing.jhtml

———, "Building the Legacy: IDEA 2004," 2004. As of August 12, 2015:
http://idea.ed.gov/explore/view/p/,root,statute,I,B,612,a,5,

Vernez, Georges, Rita Karam, Louis T. Mariano, and Christine DeMartini, *Evaluating Comprehensive School Reform Models at Scale: Focus on Implementation*, Santa Monica, Calif.: RAND Corporation, MG-546-EDU, 2006. As of March 1, 2016:
http://www.rand.org/pubs/monographs/MG546.html

Walqui, Aída, and Margaret Heritage, "Instruction for Diverse Groups of English Language Learners," paper presented at the Understanding Language Conference, Stanford, Calif., January 2012.

Walters, Kirk, Aubrey Scheopner Torres, Toni Smith, and Jennifer Ford, *Gearing Up to Teach the Common Core State Standards for Mathematics in Rural Northeast Region Schools*, Washington, D.C.: Regional Educational Laboratory Northeast and Islands, National Center for Education Evaluation and Regional Assistance, Institute of Education Sciences, U.S. Department of Education, REL 2015-031, 2014.

Wei, Ruth Chung, Linda Darling-Hammond, Alethea Andree, Nikole Richardson, and Stelios Orphanos, *Professional Learning in the Learning Profession: A Status Report on Teacher Development in the U.S. and Abroad*, technical report, Dallas, Tex.: National Staff Development Council, February 2009. As of March 1, 2016: http://learningforward.org/docs/pdf/nsdcstudytechnicalreport2009.pdf?sfvrsn=0

Wiggins, Grant P., *Educative Assessment: Designing Assessments to Inform and Improve Student Performance*, Vol. I, San Francisco: Jossey-Bass, 1998.

Wilson, Linda Dager, "High Stakes Testing in Mathematics," in Frank K. Lester, Jr., ed., *Second Handbook of Research on Mathematics Teaching and Learning*, Charlotte, N.C.: Information Age Publishing, 2007.

Wu, Hung-Hsi, "Phoenix Rising: Bringing the Common Core State Mathematics Standards to Life," *American Educator*, Vol. 35, No. 3, Fall 2011, pp. 3–13.